THE FORTY-FIRST L. RAY BUCKENDALE LECTURE

All-Wheel and Four-Wheel-Drive Vehicle Systems

SP-1063 (952600)

The L. Ray Buckendale Lectures, inaugurated in 1954, commemorate the contributions of the 1946 SAE President as developer of the latent abilities of young engineers and as an authority in the theory and practice of gearing, particularly as applied to automotive vehicles.

A cash award and certificate are provided for an annual lecture and monograph by a distinguished authority in the technical areas of commercial or military ground vehicles for either on- or off-road operation.

The Lectures are directed toward filling the needs of young engineers and students for up-to-date practical knowledge.

Sponsorship of the Lecture series is rotated among companies within the commercial vehicle industry. This Lecture has been made possible by funds provided by Eaton Corporation. The Lecture series is administrated by SAE in accordance with a Plan for Administration approved by the SAE Board of Directors.

The L. Ray Buckendale Committee of SAE welcomes nominations for future Lectures and suggestions for subject matter related to the basic objective of the program. Nominations and suggestions may be addressed to: Secretary, L. Ray Buckendale Committee at SAE Headquarters.

GLOBAL MOBILITY DATABASE

All SAE papers, standards, and selected books are abstracted and indexed in the Global Mobility Database.

Published by:
Society of Automotive Engineers, Inc.
400 Commonwealth Drive
Warrendale, PA 15096-0001
USA
Phone: (412)776-4841
Fax: (412)776-5760
November 1995

94-0130

ISBN 1-56091-613-3
SAE/SP-95/1063

Printed in USA

DEDICATION

I am dedicating this book to my wife, Bernice, who has stood beside and behind me for over 42 years. She is responsible for nearly everything good and worthwhile that I have been able to accomplish over the years.

Starting with the rigors of night school to obtain my BSME and on through career changes and many moves, her support was always there. They say "Behind every man's success there is a dedicated woman." This has certainly been the case for this grateful engineer.

Wesley M. Dick

Wesley M. Dick is Chief Engineer, Transfer Case Product, Dana Corporation. While attending night classes at the University of Toledo, Mr. Dick began his career in the automotive industry as a laboratory technician with Electric Auto-Lite Company in 1954. He joined Dana Corporation as a Test Engineer in 1957. He later was a Project Engineer in the Clutch Division, Senior Engineer in Advanced Engineering, and Section Supervisor of Engineering in the Axle Division. In 1966 he received the degree of Bachelor of Science Mechanical Engineering from the University of Toledo.

Mr. Dick joined Borg Warner Corporation in 1967 as Chief Engineer, Mechanical Products. He headed the design team that developed a family of Power Wheels (a hydraulically driven planetary drive system for agricultural machinery) and the Skid Steer Transmission for recreational all-terrain vehicles. He also participated in the development of Borg Warner's original Quadratrack Transfer Case System.

In 1972 he returned to Dana Corporation as Chief Engineer of Advanced Engineering for the Spicer Axle Division where he supervised the development of a family of two-speed heavy axles for highway trucks. This project evolved into a new major division (Spicer Heavy Axle). As Chief Engineer of Hydro-mechanical Transmissions, he worked on the development of a Continuously Variable Dual Path Hydromechanical Transmission.

Mr. Dick is the holder of 11 patents in the fields of transfer cases, transmissions, and electric actuators. He has been a member of SAE for 34 years, is a member of the Board of the Fort Wayne Section and serves as an engineer-advisor for SAE's local *World in Motion* project. He has served on the SAE National Nominating Committee, is a member of the SAE Power Train Committee, and is Chairman of the SAE All Wheel Drive Standards Committee. He has been the author or co-author of 4 SAE papers.

Wes Dick was recipient of the *Modern Metals 1988 Casting Competition Grand Award* for his role in the development of a die cast magnesium transfer case housing. In 1992 he was a co-recipient of the *SAE Henry Ford II Distinguished Award for Excellence in Automotive Engineering* for his role in the development of the Electronically Controlled Transfer Case for the Ford Aerostar.

TABLE OF CONTENTS

ABSTRACT 1

SCOPE 2

HISTORY 3

VEHICLES 9
- Commercial 9
- Military 11
- Sport Utility 11
- Specialty Vehicles 12
- Passenger Cars 12

SYSTEM COMPONENTS 14
- Transfer Case 14
  - Differentials 16
  - Differential Locking Devices 16
  - Power-Take-Offs 16
  - Reduction Gearing 17
  - Shift Devices 17
  - Synchronizers 18
  - Internal Universal Joints 19
  - Lubrication Pumps 19
  - On Demand Transfer Cases 20
- Propeller Shafts 21
  - Cardan Joint 21
    - Lead/Lag 21
    - Secondary Couple 22
  - Out-Of-Balance 23
- Front Axle 26
  - Configurations 26
    - Beam or Rigid 26
    - Independent 27
    - Twin Beam 27
  - Advantages & Disadvantages of Axle Types 28
  - Disconnects 28
- Rear Axle 30
- Half Shafts 31
- Tandem Axle As A Four-Wheel-Drive 32

DIFFERENTIAL 34
- Purpose of a Differential. 35
- Types of Planetary Differentials 38

DIFFERENTIAL LOCKING DEVICES 40
Balance Beam Analogies 40
Devices to Restrict Differentials 41
Sliding Tooth Collar 41
Sprag Clutch 42
Hydraulically Actuated Multi-Disc Clutches 42
Torque Actuated Multi-Disc Clutches 43
Electro-Magnetic Actuated Multi-Disc Clutches 44
Viscous Coupling 45
Cone Type Clutch 46
Unconventional Differential Devices 46
Torque Distribution 47
Positive Locking Devices 47
Sliding Tooth Collar 47
Sprag Clutch 47
Multi-Disc Clutches 47
Limited Slip Devices 47
Multi-Disc Clutches 47
Cone Type Clutch 47
Viscous Coupling 47
Unconventional Differential Devices 48
Placement of Clutching Devices 48
Bevel Gear Differentials 48
Planetary Gear Differentials 49

HANDLING 50
Effect Of Front to Rear Torque Balance 50
Fundamentals of Handling 51
Tracking 51
Stability 51
Cornering 51
Acceleration 51
Hydroplaning 52
Traction 52
Drivetrain Systems 52
All-Wheel-Drive 52
Part-Time 4-Wheel-Drive 52
On-Demand 53
Continuously Variable 53
Analysis Of Systems 53
All-Wheel-Drive 53
Part-time 4-Wheel-Drive 54
On-Demand 54
Continuously Variable 54

TRACTION 55
Tractive Capability 55
Tire μ vs Slip Curves 56
ABS Traction Control. 56
Difflock Traction Control. 57
Off-Roading. 57

EFFICIENCY 58
Energy Loss In Components 58
Tire Energy Losses 58
Energy management 58

TESTING 59
Laboratory Testing 59
Field Testing 59

SUMMARY AND CONCLUSIONS 60

ACKNOWLEDGEMENTS 61

BIBLIOGRAPHY 62

APPENDIX 63
1. Calculating Planetary or Epicyclic Gear Reduction Ratios . 63
2. J1952 and J2059 69

952600

ALL–WHEEL AND FOUR–WHEEL–DRIVE VEHICLE SYSTEMS

Wesley M. Dick
Dana Corporation

ABSTRACT

The purpose of this paper is three fold: to serve as a tutorial for engineers new to the field of drivetrain; to function as a reference manual for those who wish to retrieve some information on a topic they have not visited for some time; and to show the direction that four-wheel-drive technology should be taking.

A brief history of four-wheel-drive is followed by some examples showing the broad range of four-wheel-drive vehicles in use today. All components of typical systems are discussed in some detail.

The mechanics, function, purpose, and logic of different types of systems are described, as are their advantages and disadvantages. Handling characteristics and traction capabilities of some different systems are analyzed.

A majority of the vehicles produced today have systems that simply lock all drive axles together when in four-wheel-drive mode. This limits their use to off-road or extremely slippery conditions.

The mind set or focus of four-wheel-drive as merely an off-road or foul weather requirement has kept us from realizing its full potential. Full-time systems can be operated on good pavement to provide better handling, stability, safety, and comfort.

This paper will show that full-time systems should be applied to all classes of four-wheel-drive vehicles. By providing more features, improving mobility, and making better use of the technology, full-time systems offer greater value.

SCOPE

This paper gives a brief review of four-wheel-drive history to illustrate the fact that driving all four wheels is not something new and unusual. Actually four-wheel-drive was brought about by the need to achieve maximum mobility in adverse situations. The need to traverse difficult terrain and muddy roads was far greater in the early days of motorized vehicles than it is today.

There is still a need for vehicles with a pure mobility system in today's market. Examples are: mining, logging, utility work, construction work, and many recreational vehicles.

While a basic four-wheel-drive system will perform quite adequately for most situations, it will be shown that by simply adding a center differential further improvements will be obtained. A simple locking collar can be employed for those rare instances when a fully locked system is required.

A far greater number of today's four-wheel-drive vehicles are purchased by people who will rarely if ever take them off road. The purpose in purchasing a four-wheel-drive system in these sport utility and light pickup trucks is aimed at foul weather, snow, ice, etc. conditions to give some insurance of not getting stuck.

Applying torque to all of a vehicle's wheels will produce the best combination of handling characteristics, mobility, utility, safety, economy, and comfort. From the earliest concepts to modern times, the emphasis has been on pure mobility and the ability to negotiate difficult terrain. Mud, rocks, sand, steep ascents and descents, snow, and general slippery situations are certainly conditions that require driving as many wheels as possible. These situations tend to focus the application of all-wheel-drive as only required when adverse conditions exist or when the vehicle is being used for off-road operations.

Because of the long term image of four-wheel-drive as a tool to get unstuck or to prevent getting stuck, some of the most important features and benefits of these systems are overlooked and the vehicle owners in effect are cheated out of more than 90% of the potential value of a very expensive and sophisticated drive system.

Four-wheel-drive can and should be used all the time. The benefits are:

- **BETTER TRACKING** – The vehicle will maintain direction, go straight down the road for longer periods of time without steering input.
- **STABILITY** – The vehicle will have less tendency to wander and weave.
- **CORNERING** – Both oversteer, as associated with rear drive, and understeer, associated with front drive, can be nearly eliminated, thus allowing the driver to maintain a desired curve radius so as not to wander off into the berm or make contact with a guard rail.
- **ACCELERATION** – Dividing the available torque among four instead of two wheels allows the vehicle to get underway quickly without spinning out, wasting energy, and wearing away tire rubber needlessly.
- **HYDROPLANING** – Driven tires have a far greater resistance to hydroplaning than undriven tires. The four-wheel-drive vehicle can be driven much faster without loss of control and more safely at a lower speed than a two wheel drive vehicle.

Little has been written about visualization, or forming a mental image, of three dimensional systems. (Four-wheel-drive vehicles are a good example of complex three dimensional systems.) This paper shows a simplified analogical approach to this visualization process.

Bevel gear and planetary gear differentials are examined. An easily remembered tabular approach to the calculation of epicyclic gear trains is also presented.

With a more complete understanding of the systems and all the potential benefits, it is hoped that the future will see a far greater use of full-time four-wheel-drive systems.

HISTORY

FWD CORPORATION — One of the early pioneers in the four-wheel-drive field was FWD Corporation.

"Who ever heard of a mule walking on only two legs?" The year was 1906—the place was Clintonville, Wisconsin—the speaker was master blacksmith, machine shop co-owner, and soon-to-be inventor of the first four-wheel-drive vehicle designed specifically to carry passengers over rough terrain, Otto Zachow.

The concept of driving all of a vehicle's wheels goes back to the very beginning of motorized carriages. The absence of a modern road system provided very challenging mobility situations. Ruts and deep mud holes were the order of the day. It soon become very obvious that driving only one or two wheels was not adequate to negotiate these extremely difficult conditions.

Otto Zachow and his partner (also brother-in-law), William Besserdich, operated a small machine shop where they repaired steam engines and farm equipment and sold Reo cars.

What Otto didn't know when he began work on his system, was that there already had been over 700 patents issued on attempts to accomplish this very thing. Reputedly inspired by the tumbling rods of a threshing machine, he conceived a mechanism that would deliver power to all four wheels. He then enlisted the aid of his son, who was apprenticed as a machinist and draftsman, to help him put the concept on paper. With the advice of a local attorney, Walter A. Owen, application was made for a patent (the first of several) which was granted March 24, 1908.

The Zachow-Besserdich machine shop then began construction of a car using Otto's ideas and the partners devoted all their spare time to the project. The chassis of that first car, a steamer, was unveiled December 26, 1908. (*Figure 1*)

Figure 1
FWD's 1st steam powered 4WD vehicle

The local newspaper reported that it reached a speed of 25 mph, plowed through snowdrifts three feet deep, went over rough and slippery roads, and climbed the steepest hills.

Now that their basic car had proved successful, more funds were needed to complete it. In May a large 8 passenger body was ordered on credit, a Dr. Finney became a 3rd partner, and $35,000 was raised to fund the Badger Four Wheel Drive Auto Company.

Progress was slow. By 1910 the steam car still was not completed and investors were disgruntled. It was then Zachow turned once again to his lawyer, Walter Owen. Owen secured more capital from community businessmen, and the company was reorganized as the Clintonville Four Wheel Drive Auto Company.

Pleased with the car but unhappy with engine performance, the partners rebuilt their four-wheel-drive auto and powered it with a four cylinder gasoline engine. They named this vehicle the "Battleship." (*Figure 2*)

Figure 2
FWD's Gas Powered "Battleship"

Its drive train was similar to many four-wheel-drive systems in use today. There was a transfer case mounted behind the transmission. The rear axle drive was straight through and the front drive was through a chain that offset the output to the front axle. There was no differential in the transfer case and no provision for disconnecting the front drive system. Since the roads of the day were mostly mud, sand or gravel the lack of an inter-axle differential or disconnect was not a serious issue.

In 1912 FWD began producing trucks for the U.S. Army with basically the same drivetrain as the earlier passenger car. They later incorporated an inter-axle differential that was lockable, thus making a major step in the four-wheel-drive technology. FWD would go on to produce over 20,000 four-wheel-drive trucks for military use in World War I. The FWD Company continues to make four-wheel-drive vehicles to this day.

Unfortunately Otto Zachow did not participate in FWD's longevity. In 1910 he resigned to spend more time on his machine shop business. The inventor of the double-Y universal joint and the ball and socket drive system sold all his rights and interests in the FWD Corporation before his design became successful. In 1914 Besserdich also sold his interest in FWD. Walter Owen, the attorney who assisted in securing the initial patent and reorganized the company, stayed on to serve as company president for more than 40 years.

EARLIER VEHICLES — There were at least three four-wheel-drive vehicles produced before the FWD car, but none relates to modern four-wheel-drive designs. They were:

The Columbia Electric Truck — Colonel Albert Pope built electric cars in his Columbia Automobile Co. plant in Hartford, Connecticut beginning in 1897. In 1901 the name was changed to the Electric Vehicle Co. In 1902 they built the Columbia Electric Truck, a 5-ton monster powered by four electric motors, one on each of the four chain driven wheels. The Columbia had the first power steering system in the form of a fifth wheel mounted parallel to the ground, with the front-mounted electric motors geared into it. It was not intended for off-road use and the four-wheel-drive was just one benefit of use of electric drive. Records show that approximately 8,000 Columbia trucks were in use around 1907; however not all were four-wheel-drive and there is no record of the number of four-wheel-drive Columbia trucks built.

The Aultman Steam Wagon — The steam powered Aultman truck was built in Canton, Ohio between 1901 and 1902. Apparently only one FWD unit was ever completed. This was a 5-ton truck with a 16 hp, 2 cylinder motor and a friction transmission. The four-wheel-drive system had double chains to the rear axle and bevel gears to the front axles. (One problem with designing four-wheel-drive vehicles is how to turn and steer them. In the Aultman this problem was solved by a double front axle: (1) one the weight bearing axle with a normal steering knuckle and (2) the drive axle, mounted above and parallel to the first axle.)

The Cotta Car — A much lighter vehicle classed as a motorcar was the Cotta, built in Rockford, Illinois. The four-wheel-drive system consisted of four separate chains connected to a differential gear in the center of the chassis. It also had a flexible U-joint. At least one car was completed. The company was later purchased by FWD.

JEFFREY QUAD — Another early manufacturer of four-wheel-drive vehicles was Thomas B. Jeffrey, an English-born inventor who in 1879 began manufacturing the Rambler bicycle in Chicago. Later he switched his interest to automobiles, and purchased the Sterling Bicycle Company's old factory in Kenosha, Wisconsin to manufacture them. In 1902 he and his son were making Rambler automobiles there. The company prospered, and in 1914 the name "Jeffrey" replaced the Rambler trademark. In 1917 Charles W. Nash, former president of GM, purchased the company and it operated as Nash Motors until 1954, when it became the Hudson Motor Car Company. Eventually this company merged with American Motors Corp., which was finally purchased by Chrysler.

In 1913 when the Army visited various truck manufacturers, hoping to find something to replace the mule, they were impressed by FWD's four-wheel-drive vehicles, but not by the company, which had only made 24 trucks to date. On the other hand, Jeffrey had built tens of thousands of cars and hundreds of trucks, none of them four-wheel-drive.

Since Jeffrey had much more to offer than FWD in terms of manufacturing facilities and experience in building trucks, the Army asked Jeffrey engineers to design a 1½ ton 4x4. The truck they designed was called the "Jeffrey Quadruple-drive truck" — soon shortened to the "Jeffrey Quad." While similar in general appearance to the FWD, the "Jeffrey Quad" had several distinct advantages. It was powered by a 36 hp Buda four-cylinder engine and used a four speed transmission. It used four wheel steering, which allowed a turning circle of just 45 feet, and it had four wheel brakes.

In the first year of production Jeffrey built 5,578 trucks. (Most were four-wheel-drive.) Both the rival FWD vehicle and the Jeffrey Quad passed Army tests and soon were to be found around the world. Jeffrey Quads accompanied General Pershing's army to Mexico and later to Europe. The need for Quads was so great in World War I that the Jeffrey factory couldn't meet the demands and other manufacturers (Hudson, National, Paige, etc.) built Quads under Nash license and patent.

The Quad was built from 1913 to 1928. During that time 14,674 units were built and sold. In addition to Army vehicles there were Quad armored trucks, stake trucks, dump trucks, express trucks and fire trucks. After the war, surplus 4x4 Quads went to work for state highway departments and the company, now owned by Nash, began converting production lines to passenger car manufacturing Since Jeffrey/Nash did not develop a four-wheel-drive parts and service department, many Quad trucks had a shorter life than those made by FWD.

During the twenties and thirties use and development of four-wheel-drive vehicles continued. Snow removal and other government operated services were a good market for vehicles that could provide maximum traction. *Figure 3* shows a V - blade plow mounted on an FWD vehicle.

Heavy haulers such as the logging industry also found four-wheel-drive vehicles to be extremely valuable with their ability to go almost anywhere. *Figure 4* shows a 1924 version of a four-wheel-drive tractor pulling a large logging trailer. Bigger and more powerful versions of the four-wheel-drive rigs were produced as can be seen in *Figure 5*, which shows a 1978 FWD with a mounted single direction snow blade.

A 1979 concrete mixer chassis shows off an interesting 8 x 8 or four axle truck with all wheels driven. This vehicle is shown in *Figure 6*. Some very large and rugged all-wheel-drive (AWD) vehicles have been produced such as this Class 8 (GVWR over 30,000 KG) - 6 x 6 built by Oshkosh in 1977. (*Figure 7*)

Figure 3
V Blade Snow Plow

Figure 4
Log Trailer Tractor

Figure 5
Single Direction Snow Plow

Figure 6
Concrete Mixer Chassis

Four-wheel-drive vehicles have continued to evolve and will no doubt do so long into the future. As with any technology the quantum leaps were made early on, then refinements continued to be discovered and implemented. Modern materials will make the systems stronger and lighter. The application of electronics can continue to improve function and ease of operation for the driver.

We have seen the use of all-wheel-drive span passenger cars, trucks, utility and farm vehicles. Some of the applications will be discussed in the next section on vehicles.

Figure 7
Class Eight 6 x 6 Tractor

VEHICLES

Four wheel and all-wheel-drive systems have been used in all types of vehicles for both commercial and personal use.

COMMERCIAL VEHICLES — Four-wheel-drive Systems in commercial vehicles range from Class 8 trucks all the way down to light pickup trucks. There are also quite a number of construction, mining. farming, and utility vehicles of all types

Trucks — Probably the earliest use of four-wheel-drive was in delivery trucks that needed the mobility to make deliveries with the unimproved muddy and rutted roads of those early times. FWD corporation was one of the early pioneers of the four-wheel-drive truck and they are still in the business of producing a line of mixer trucks and other heavy utility type bodies. One of the very successful applications of all-wheel-drive systems today is the ready mix concrete truck. A typical chassis layout is shown in *Figure 8-a.* The chassis shown has a remote (frame mounted) transfer case, and standard drive shaft to transmit power to the front and rear axles. Only one rear axle is shown but typically multiple rears are used. *Figure 8-b* is a photograph of a typical mixer truck from 1955. Today's mixers use the same format but have multiple rear axles to carry larger mixers. (*Figure 6 under HISTORY*)

The larger series of four-wheel-drive pick-up trucks are extensively used by utility companies, construction companies and many western farmers. The smaller four-wheel-drive pickups are very popular with the rugged individualist who wants to be able to haul material for home projects, or go camping, hunting, and fishing in remote regions.

Farm Equipment — Four-wheel-drive farm tractors have become very popular in the last 20 to 25 years. The size and power of drawbar type vehicles increased as farms became larger. Examples of these heavy duty four-wheel-drive tractors are those built by Steiger and Versatile. Four-wheel-drive harvesting and other farm utility vehicles have also become more common in recent years.

The Hagie Sprayer is an interesting application of four-wheel-drive technology. This vehicle has a hydraulic drive system. Its engine runs a variable displacement pump that supplies fluid under pressure to each of the four wheels through a common conduit. The wheels have fixed displacement motors that drive the wheels through a planetary reduction wheel drive. Differential action is achieved by each wheel motor drawing fluid from the supply as required by its speed of rotation.

There is no device for restricting differentiation, but very good mobility is achieved just by having all the wheels driven. There is a good lesson here. This is a vehicle intended for off-road use (spraying crops) and traction enhancement over and above four-wheel-drive is not found necessary.

Being a farm type vehicle, the basic design working speed is set for its working mode (0 to 14 mph). To obtain a more useful speed range for transport between jobs the fluid can be rerouted so it flows through the wheels of just one axle. This results in the effective volume flow to each wheel being greater. Drawbar capability is reduced but it is not required for transport. The speed range of the vehicle for on-road transport is increased to 20 mph.

Construction — This industry has also found uses for four-wheel-drive systems. Material handling vehicles such as fork lift trucks for building sites are good examples of the technology.

Mining — Open pit mines use huge four-wheel-drive dump trucks to transport both overburden and ore. Large front end loaders using four-wheel-drive can load the biggest of on highway dump trucks in just one scoop. Under-ground mines also use a variety of four-wheel-drive vehicles for moving material and people.

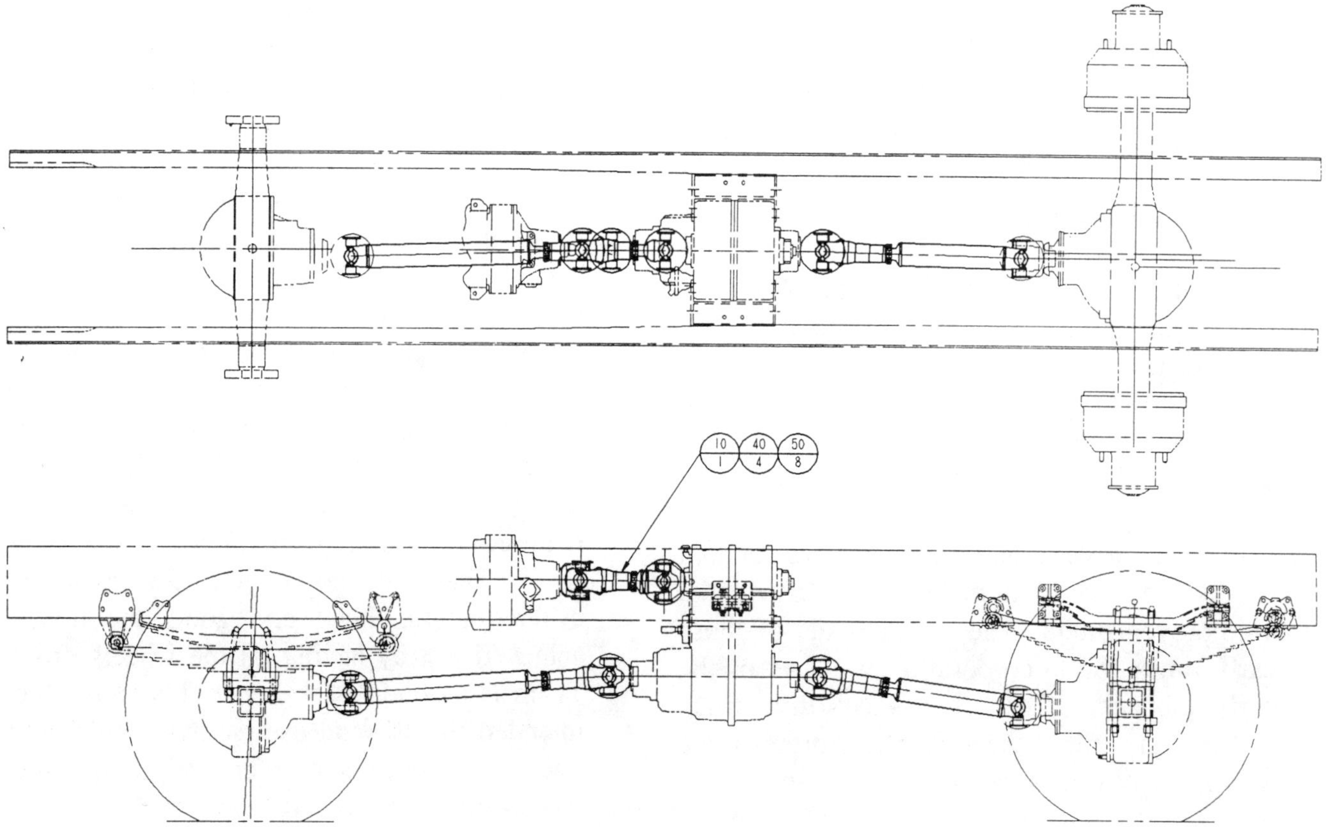

Figure 8-a
Typical Four-Wheel-Drive Truck Chassis

Figure 8-b
Typical Four-Wheel-Drive Mixer From 1955

MILITARY VEHICLES — The Mexican war created a need for many four-wheel-drive vehicles for transporting troops and supplies over the rugged terrain and unimproved roads in that area.

World War I and World War II both gave a boost to the all-wheel-drive technology since many military operations involve negotiating bombed out roads, unimproved roads and a great deal of off-roading.

World War II also was responsible for the creation of small four-wheel-drive vehicles, especially the Jeep, which was produced in very large numbers and created the awareness of four-wheel-drive among the general public.

Most military vehicles are four-wheel-drive or all-wheel-drive. These range from the small scout vehicles like the Jeep, through the now popular Hummer. Also there are rubber-tired armored weapons vehicles, six by six troop or supply trucks, and heavy duty tank rescue flatbeds.

SPORT UTILITY VEHICLES — Sport Utility Vehicles such as the Jeep Grand Cherokee, Ford Explorer, GM Blazer, Land Rover and Isuzu Trooper have enjoyed a fantastic increase in popularity in the last few years. One reason is the desire to own a four-wheel-drive vehicle. Most buyers see the four-wheel-drive option as a means of increasing mobility, especially in foul weather conditions such as snow, ice etc. There is a certain comfort in knowing that you can go virtually anywhere anytime.

These vehicles have been appointed with the finest of interiors and are equipped with all the latest features you can find on some of the most expensive luxury sedans. For this reason they are accepted at the country club as well as at the hunting lodge or the campground. Again the foul weather capability is a further enhancement to the desirability of this type vehicle.

Strangely enough a very small percent of these vehicles are used for any serious off-roading, even though they are engineered and tested to provide this capability.

Off-road trails are available in many parts of the country. Jeep holds rallies to promote the off-roading fun offered by this type vehicle.

The Rubicon trail, located in California, runs from Georgetown to Lake Tahoe and is one of the more famous of the off-road tests of driver skill and the ability of the vehicles. It is very difficult (rated 10) and offers a challenge to the most experienced off-road enthusiasts.

The area around Telluride, Colorado offers old mining trails that are steep, rocky, dangerous, difficult, and offer breathtaking scenery to those that want to get the most from their four-wheel-drive vehicles. This is an adventure that tries both the vehicle and the operator.

Jeep offers a full-time system, containing an inter axle differential and a choice of traction enhancement by either a viscous coupling or a positive lock across the differential. Some of their systems can be run in two-wheel-drive, four-wheel-drive with a differential, locked four-wheel-drive and a low ratio for speed control when off-roading.

The Ford Explorer offers an on demand system that has no inter-axle differential, but provides torque to the front axle through a multi-disc clutch when a rear wheel spins out. It can also be operated in two-wheel-drive, locked four-wheel-drive, and has a low ratio for slow speed off-road operation.

General Motors, Toyota, Rover, Isuzu and others also offer versions of Sport Utility Vehicles.

SPECIALTY VEHICLES — Some other interesting four-wheel-drive vehicles have been built.

Precursor to Jeep — *Figure 9* is a picture of the Bantam vehicle. While this was not an all-wheel-drive vehicle, it was the basic chassis that was modified to four-wheel-drive and was to become the famous WW II Jeep.

Figure 9
Bantam (Precursor to Jeep)

Race Cars — Four-wheel-drive race cars have been produced and some have run quite successfully. The Granitelli turbine car was one of the most famous. It was so far advanced over the competition that the rules were changed to prevent further running of that type of car. In the days before the modern sophistication of aerodynamics to hold the cars on the turns it was said the four-wheel-drive capability was as much a contributor to the cars extraordinary performance as the turbine engine.

Two earlier four-wheel-drive racers were built. One by Harry Miller of FWD, for the 1932 Indianapolis 500 (*Figure 10*), and the FWD Special that ran at Indianapolis in 1933, 1934, and 1936 (*Figure 11*).

Figure 10
Harry Miller Four-Wheel-Drive Racer

Figure 11
FWD Four-Wheel-Drive Racer

PASSENGER CARS AND MINIVANS — Four-wheel-drive passenger vehicles have never been built in abundance. A few that have been produced in modern times and their four-wheel-drive systems will be described.

Subaru — The Subaru is produced in two four-wheel-drive versions. One has a manual transmission and a full-time transfer case containing a center differential and a viscous coupling to restrict differentiation when wheel slip occurs. The other is an automatic transmission version that drives the second axle through a viscous coupling. This is basically an on demand system that will only deliver torque to the second axle in the event of a wheel spin out on the primary driving axle.

Ford Scorpio — The Scorpio had a full-time system with a viscous coupling across the inter axle differential. Its torque split was 60% rear, 40% front. A drop chain drive provided power to the front axle.

BMW 525iX — The BMW 525iX has a longitudinal engine with a conventional transfer case layout. The inter-axle differential is a planetary with 64% of the torque to the rear and 36% to the front. It has an electromagnetic multi-disc clutch acting between the sun gear and the ring gear of the differential. Actuation of the clutch is by computer control to enhance mobility. Offset to the front drive is by chain. The rear axle contains an electro-hydraulically actuated limited slip device for further traction enhancement.

Aerostar — The Ford Aerostar has one of the most sophisticated and effective traction enhancement systems in production at this time. It is a full-time system with a planetary differential that normally splits torque 33.3% to the front and 66.6% to the rear. Hall effect sensors count teeth on the front and rear outputs. An onboard computer analyzes the relative movement of the front and rear, then compares with stored data to determine if there is normal differentiation or if a wheel slip is occurring at either axle. In the event of a wheel slip or spin out, it engages an electromagnetic clutch so additional torque can be applied to the non slipping axle.

Safari/Astro — The GM minivan has a similar system with a 40% front to 60% rear torque split. Traction enhancement is obtained by employing a viscous coupling across the differential. When a wheel spin occurs, some additional torque is transferred to the opposite axle and if the wheel spins faster a greater amount of torque will be transferred.

Voyager/Caravan — The Chrysler minivan four-wheel-drive is based on a front drive passenger car chassis. It uses a power take-off approach to drive the rear axle. There is no inter-axle differential and a continuously rotating rear propshaft is attached to the rear axle via a viscous coupling. In cornering when the front is slightly overrunning the rear a small amount of torque will be transferred to the rear. When there is a front spin out, torque will be applied to the rear in proportion to the speed of the spinning wheel.

SYSTEM COMPONENTS

The basic drivetrain components comprising four-wheel-drive systems are: a transfer case, propeller shafts, front axle, rear axle and half shafts. Use of a tandem axle will also be discussed.

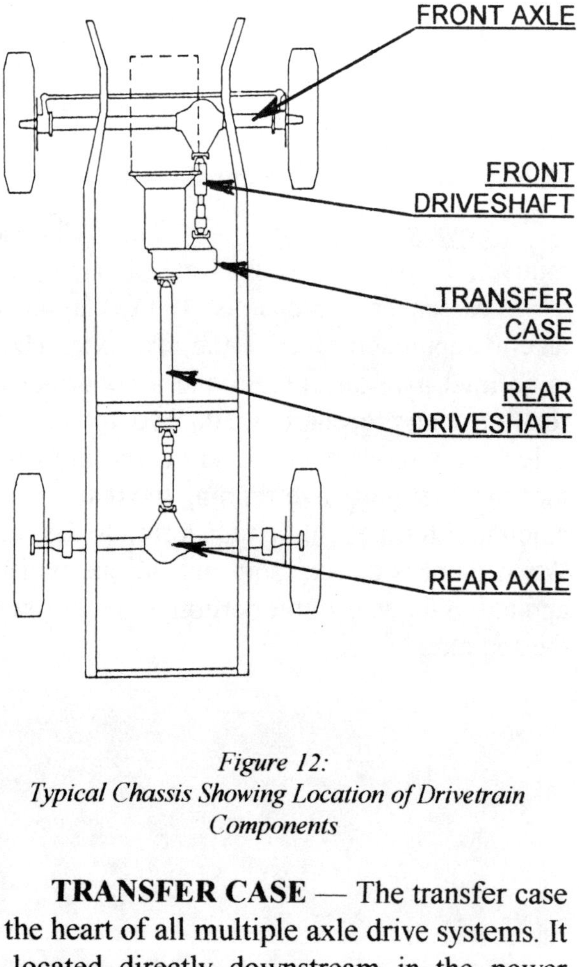

Figure 12:
Typical Chassis Showing Location of Drivetrain Components

TRANSFER CASE — The transfer case is the heart of all multiple axle drive systems. It is located directly downstream in the power train from the transmission. In some cases it is even adapted as a permanent part of the transmission. The function of a transfer case is to direct engine power to the appropriate or desired axle or axles.

Components that <u>may</u> be found in a transfer case are: an additional gear set for low speed operation; a differential: offset gearing or a chain for front drive; front disconnect; a differential lock or clutching device; and possibly a power-take-off.

Earlier transfer cases were very simple and usually served two functions: (A) To connect power to the front axle for off-road operation and (B) to provide an additional low gear ratio so the vehicle could be driven at the low speed necessary for negotiating extremely rugged terrain. This is basically the system developed for the World War II Jeep.

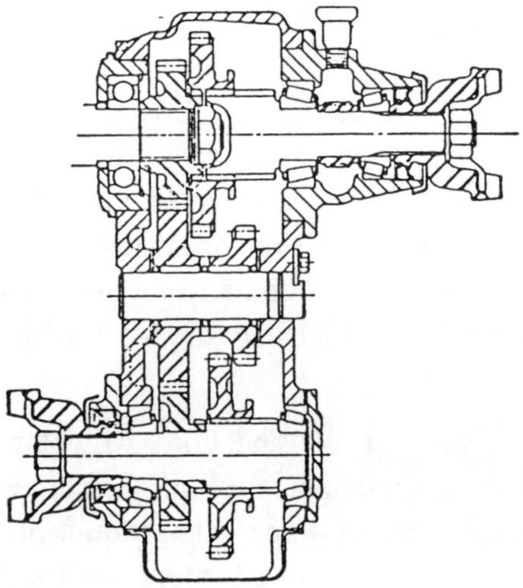

Figure 13
Sliding Gear Transfer Case as Used in WW II Jeep

Cross sections of several other types of transfer cases follow:

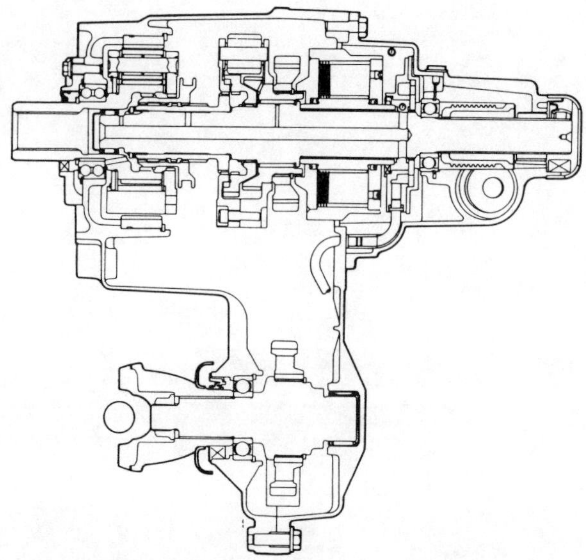

Figure 14
Full-Time 2-Speed Transfer Case with Viscous Clutching Device Across Differential for Traction Enhancement

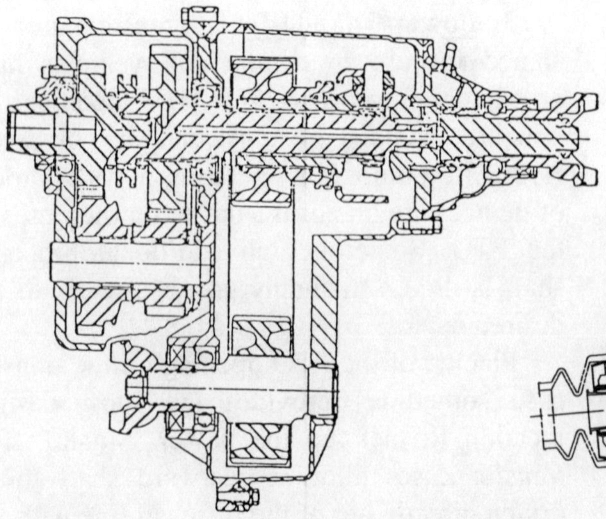

Figure 15
Full-Time Transfer Case with Differential, Positive Differential Lock, and Counter Shaft 2-Speed Arrangement

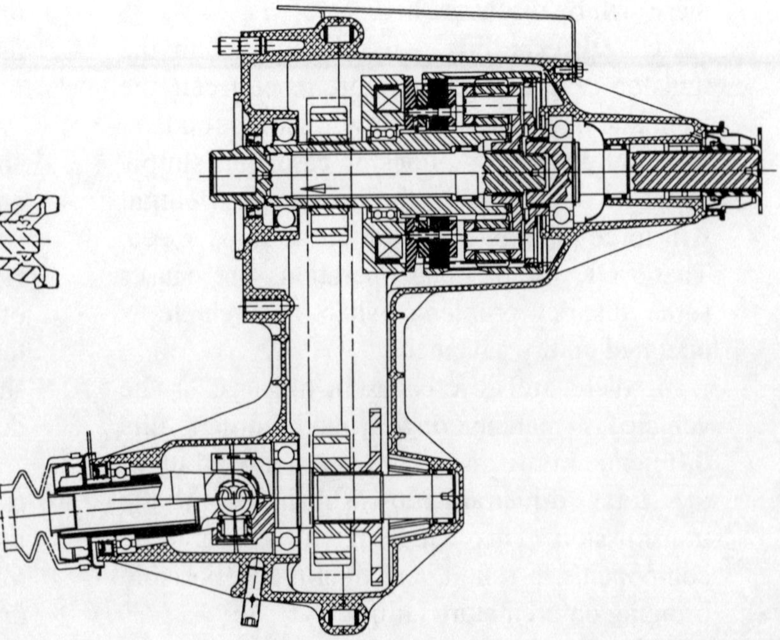

Figure 17
Full-Time Transfer Case with Differential, Electronic Traction, and Angled Output

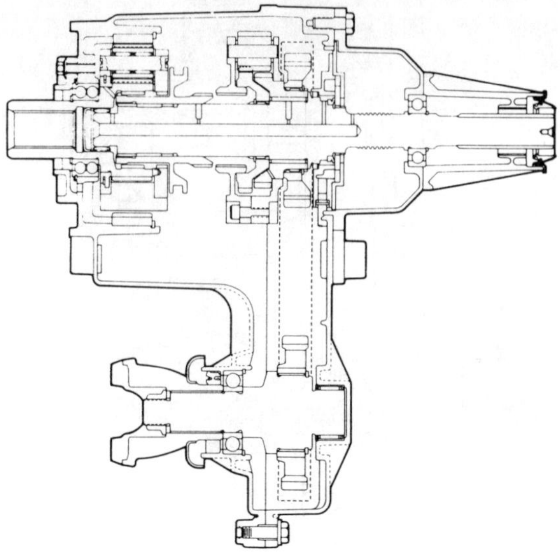

Figure 16
2-Speed Part-Time, Full-Time with Dog Clutch to Lock Interaxle Differential

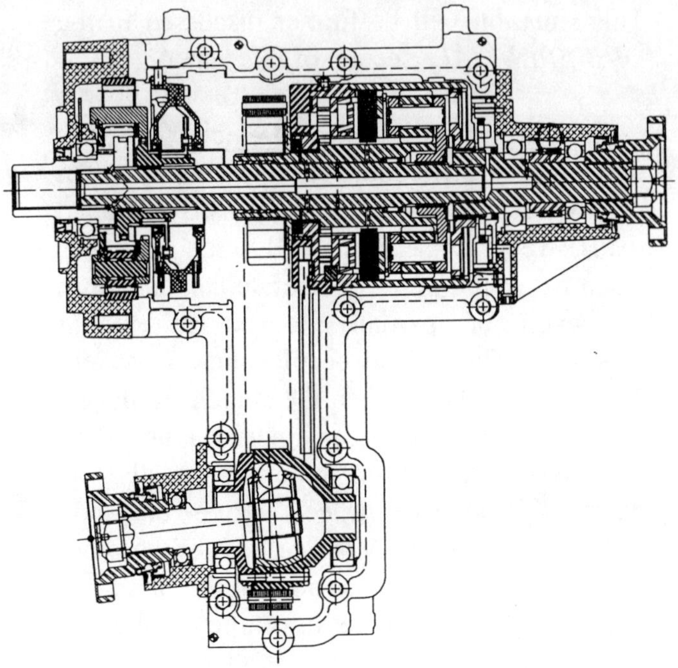

Figure 18
Full-Time Transfer Case with Hydraulic Multi-Disc Clutch for Traction Enhancement, Vacuum Shift 2-Speed, and Angled Output

Many features have been incorporated into transfer cases over the years. Some of these will be discussed here:

1. **Differentials** — A differential in the transfer case divides the torque between the front and rear axles and allows them to rotate at different speeds. A transfer case that simply connects both axles to the transmission output will force both axles to turn at the same speed. That's OK for off-road operations but causes some distinct problems when the vehicle is operated on dry pavement.

If there are differences in tire size or the vehicle is maneuvering around corners, the difference in tire speed will cause a wind up of the total drivetrain downstream from the transmission. This wind up will cause these components to run at very high loads, possibly bringing on premature failure.

Also the wind up condition will cause the wheels to frequently skid out as the wind up torque is relieved. This is especially annoying in cornering as the vehicle will tend to "Hop" around the corner and the steering wheel will also be jerked around in the driver's hands. This situation will be further discussed in the *DIFFERENTIALS* section of this paper.

2. **Differential Locking Devices** — Now that we have differentials to allow us to run in All-wheel-drive mode on the hard surfaced roads, we find there are still some instances when we would prefer to operate the vehicle in a locked or partially locked differential condition. This would be of course in severe off-road conditions when the vehicle could get hung up, such as getting in a position where the vehicle's weight is supported mostly by diagonally opposite wheels. Here the axle differentials will allow one or both of the other wheels to spin out, and as a result the same low amount of torque will be transmitted to the two wheels supporting the weight of the vehicle. Locking out the differentials will overcome this condition and the vehicle can drive out of the entrapped situation.

Many types of differential locking devices have been devised and a number will be discussed in the *DIFFERENTIAL LOCKING DEVICES* section.

3. **Power-Take-Offs** — Transfer cases designed for use in commercial vehicles have employed power-take-off (PTO) openings so onboard vehicle systems can be operated. Hydraulic pumps for operating a wide variety of devices are one of the most common uses of the PTO. Sometimes an additional propeller shaft is driven to supply power directly to the desired device.

Placing of the PTO opening on the transfer case sometimes provides the most desired location to pick up the power. Single offset transfer cases allow a rearward shaft to be driven directly out of the offset in line with the front driveshaft. The necessary internal clutching collars must of course be designed in to direct power to the axle propshafts or the PTO as desired by the vehicle operator.

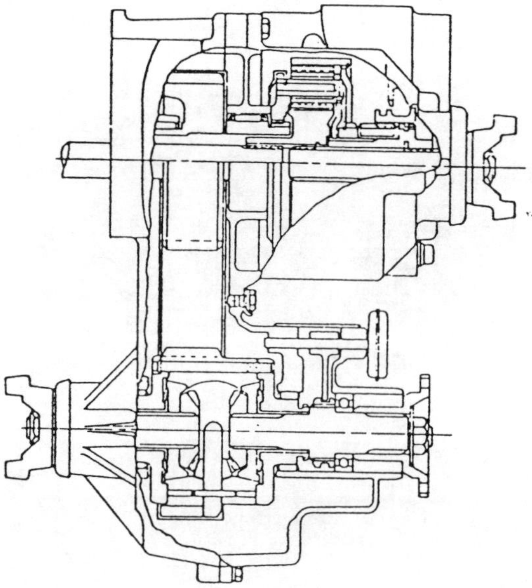

Figure 19
One Arrangement of a Transfer Case with Power-Take-Off

4. **Reduction Gearing** — Early transfer cases and current heavy duty transfer cases usually use countershaft type gearing when multiple ratios are required. (*Figure 20*)

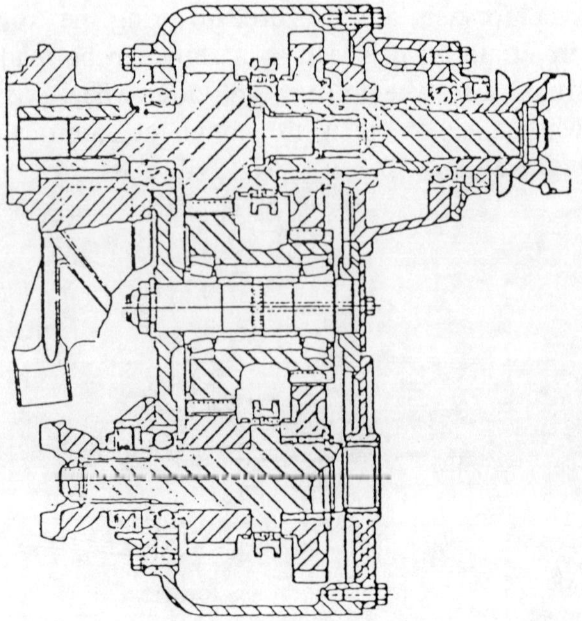

Figure 20
Early Countershaft Arrangement

The high volume transfer cases used in sport utility and light pickup trucks mostly use a planetary gear set for the reduction of output speed. The gearset can also be used as a speed increaser. This gearset is normally placed in the front of the transfer case. The input shaft is connected directly to the sun gear and the ring gear is mounted stationary in the case. With this arrangement the carrier becomes the reduction output and direct (1 to 1) ratio is accomplished by connecting the output shaft to the sun gear. The shift collar alternately connects with the carrier for reduction and with the sun for direct. An example of this arrangement is shown in *Figure 21.* Other arrangements of planetary gearsets are possible and are or have been used

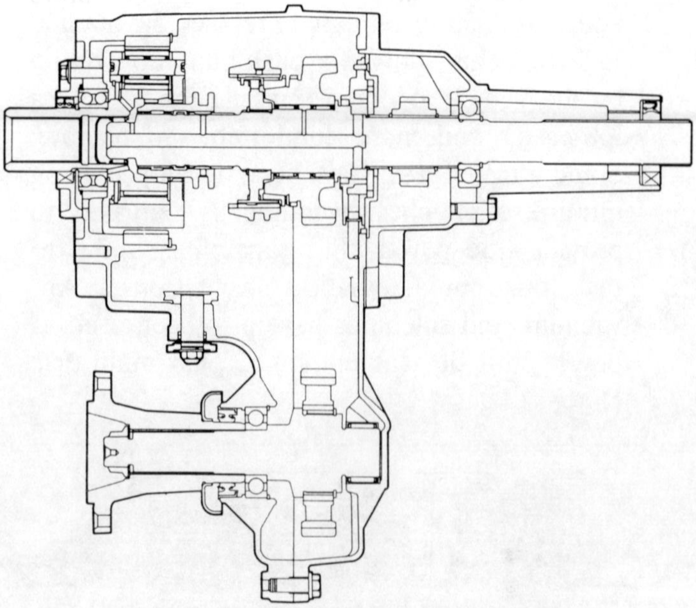

Figure 21
Transfer Case with Planetary Reduction in Front Section

Calculating planetary gear ratios is an interesting topic in itself. There are books of formulas for a multitude of planetary arrangements. To use these, you identify the arrangement and plug into the appropriate formula. The problem with this is that the derivation of the formula may not be fully understood and, as a result, errors may be made. In the appendix you will find an excerpt from *Guillet's Kinematics of Machines* showing a tabular method of calculating planetary or epicyclic gear ratios. This is from a textbook I studied during my college days. In all the years since, I have never found a better system for calculating epicyclic gear ratios.

5. **Shift Devices** — A wide range of shifting devices have been used. In the older gear driven boxes, both sliding gears and sliding collars moved via shift forks were common. In modern chain driven boxes, sliding collars moved by shift forks are still quite common. Manual levers and shift linkages are attached to the fork and collar mechanisms. There are also quite a number of

applications employing electric motor devices added to the transfer case to replace the manual linkage. This allows for push button operation by the vehicle operator, making the vehicle a lot easier and more comfortable to operate. Some transfer cases have been developed using multi-disc clutches hydraulically actuated to provide true power shift capability similar to that found in automatic transmissions. Air, vacuum, and solenoids have also been used as power shift devices employed with multi-disc clutches.

Some vehicles are set up with front disconnect devices such as hub locks so the front drive train can be allowed to stop during two-wheel-drive operation. The reasoning here is to squeeze out better fuel mileage by eliminating the energy loss associated by spinning these components. For this reason synchronizers are also used to bring the front drivetrain up to speed so a shift can be made from two-wheel-drive to four-wheel-drive while the vehicle is in motion at significant highway speeds.

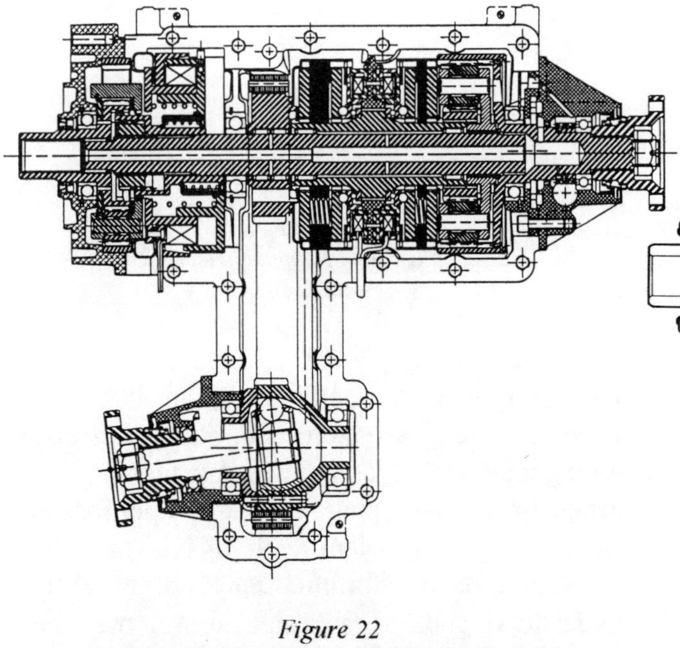

Figure 22
Transfer Case with Solenoid Shift Low to High, Power Shift 2-Wheel Drive to 4-Wheel-Drive, Electronic Clutch for Traction Enhancement, and Constant Velocity Joint for Angled Output

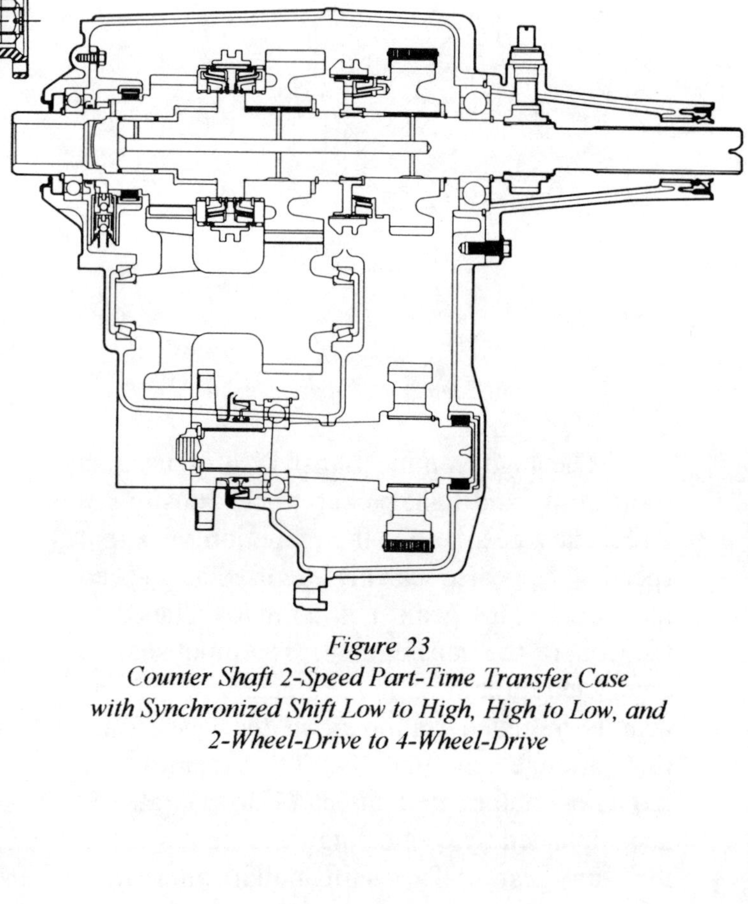

Figure 23
Counter Shaft 2-Speed Part-Time Transfer Case with Synchronized Shift Low to High, High to Low, and 2-Wheel-Drive to 4-Wheel-Drive

6. **Synchronizers** — Some transfer cases have been designed with synchronizers in their shift mechanisms. The purpose is to allow shifting of low to high range or two to four wheel operation while the vehicle is in motion. Transfer case low range ratios are usually very deep. 2.5 to 1 up to 2.8 to 1 are normal and some are even greater. For this reason there is a very large excursion in engine speed when making a low to high or high to low shift. The shift is very difficult but can be accomplished if a rugged enough synchronizer is used.

7. **Internal Universal Joints** — Internal universal joints have been employed to make propeller shaft angles within an acceptable range. The most common location for a universal joint has been at the front output, where the most severe propshaft angles are usually encountered. Either standard cardan joints or constant velocity (CV) joints can be used. They can be employed at either output and/or at the transmission input.

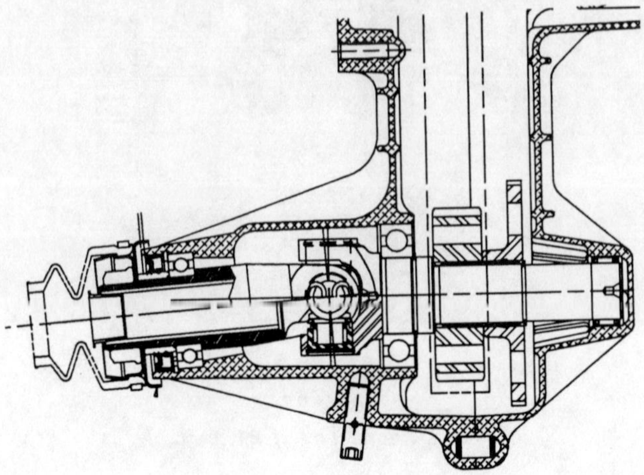

Figure 24
Cardan Joint for Angled Output

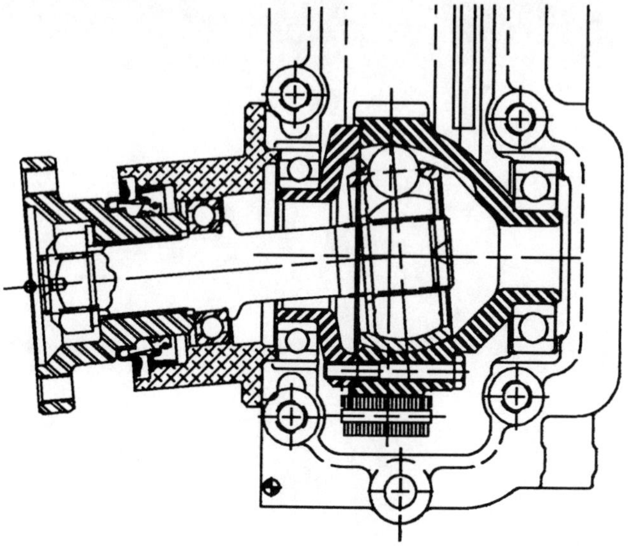

Figure 25
CV Joint For Angled Output

8. **Lubrication Pumps** — Lubrication pumps are common in modern production transfer cases. Transfer cases having planetary reduction units generally have a pump to supply oil to the fast running planet gears. Rotational speeds in excess of 20,000 rpm are common in these units. The planets are set on needle bearings and adequate lubrication is required for cooling and to keep the bearings running properly. Small piston pumps have been used, but there is now a trend toward using the gerotor type (internal/external) gear units. These are simple, economical to manufacture and provide more than adequate pressure for lubrication purposes. Pumps are sized to provide adequate flow without pumping an excessive amount of oil and therefore wasting energy.

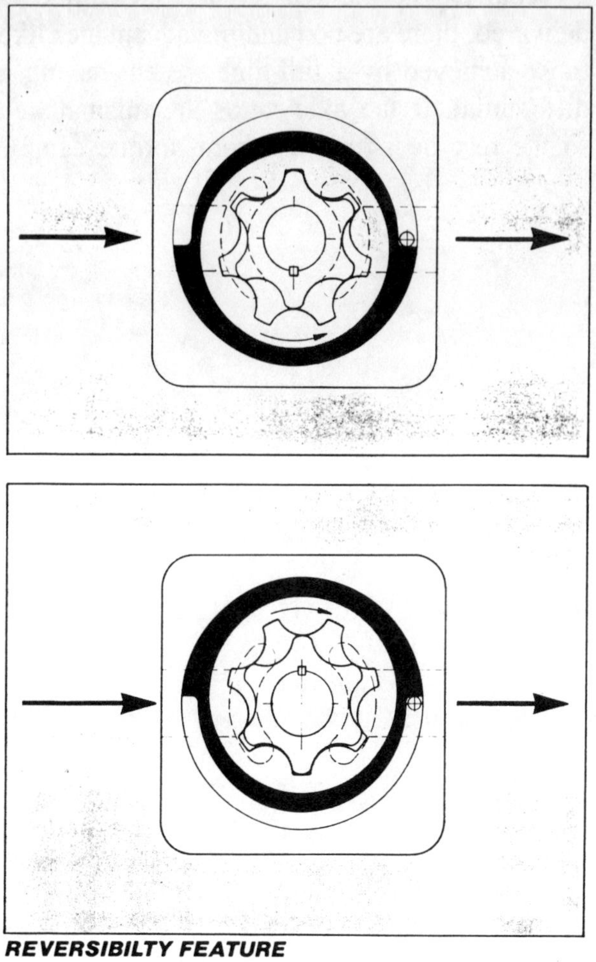

Figure 26
Gerotor Type Lubricating Pump

9. **On-Demand Transfer Cases** — This type of transfer case has no inter-axle differential. It attempts to simulate full-time four-wheel-drive by modulating a clutch that delivers torque to the front axle. The premise for the arrangement is that some gain in fuel economy will be achieved by running the vehicle with the front drivetrain disconnected between the wheels and the transfer case. This will eliminate some stirring action in the front axle, some stirring action in the transfer case if the chain and sprockets are stopped, and the associated bearing and seal drag of the front drivetrain components.

It leaves the designer with a functional dilemma. Since the front wheels normally over-run the rear in a vehicle with matched axle ratios, front to rear, the only time torque can be delivered to the front is when a spin out is occurring at the rear. When no torque is delivered, there are no handling advantages like those achieved by a full-time system having a differential. If the axle ratios are mismatched so the rear over-runs the front, torque can be delivered to the front but the clutch will be slipping any time it is used unless total lock is achieved. In either case efficiency is sacrificed, thereby detracting from the possible fuel savings the system might offer.

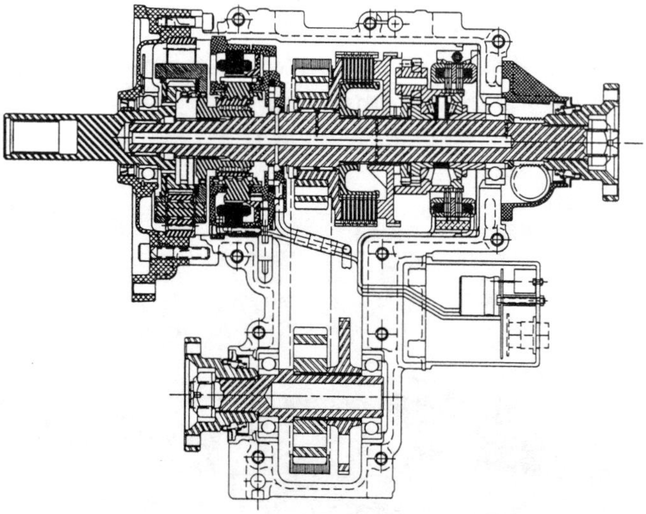

Figure 27
On-Demand Transfer Case

PROPELLER SHAFTS — The two most popular propeller shafts are those with cardan joints and those with Rzeppa (shê′ pa) type ball joints. (*Figure 28*) Tripot (or tripod) joints, along with a plunging 2 trunnion joint, are also frequently found.

Since the cardan type is by far the most common, some of its characteristics will be discussed. For a complete analysis of cardan and other types of joints, refer to SAE Publication AE-7.

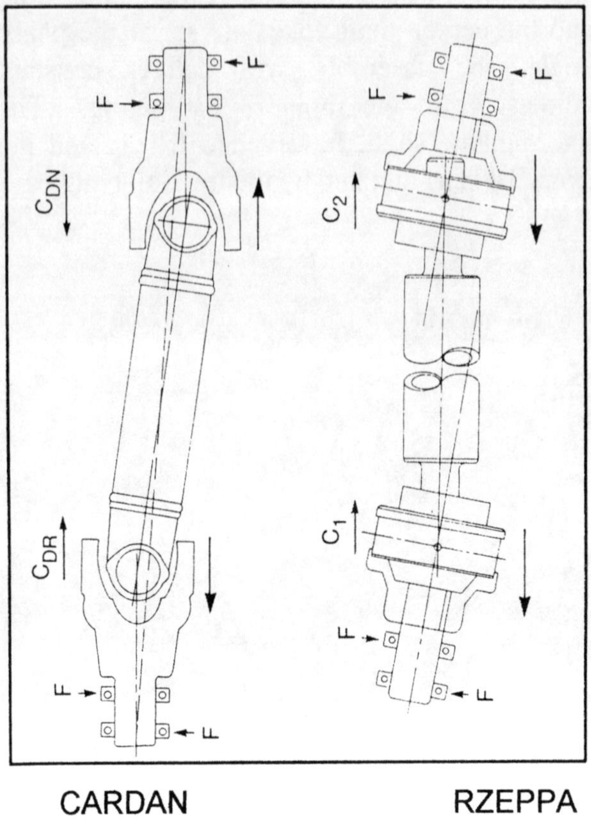

Figure 28
Cardan and Ball Type Joint

Cardan Joint — This is an introduction to the dynamics of cardan joint drive shafts.

There are three vibration forcing functions of a cardan joint. They are:

(a) **Lead/Lag of Single Joint** — When the joint is running at an angle with one yoke turning at a constant rate the other yoke turns at a variable rate, tending to go ahead and get behind the constant rate yoke. The amount of lead and lag is a function of the joint operating angle, and increases dramatically with increases in the operating angle. The lead/lag function occurs twice per revolution. Taken from a neutral position, the variable yoke will lead for 90°, lag for 90°, then lead again. This occurs as a sinusoidal function. (*Figure 29-a* and *29-b*)

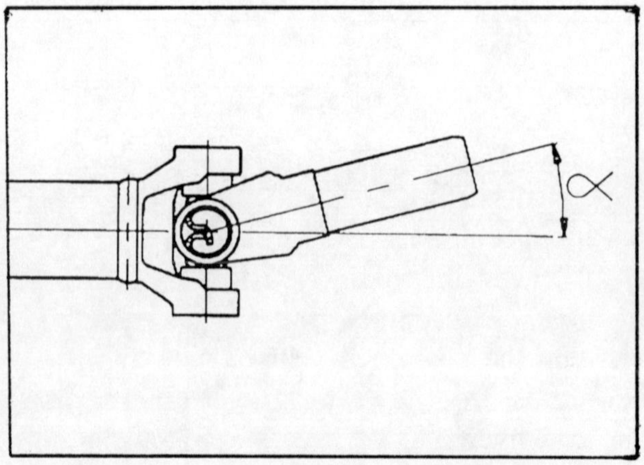

Figure 29-a
Joint at 30° Angle

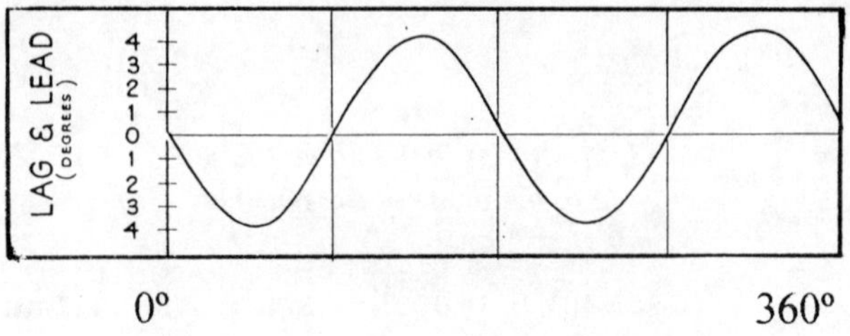

Figure 29-b
Lead/Lag Curve of 30° Operating Joint Angle

This function would, given constant engine speed, try to increase wheel speed, then decrease wheel speed twice per revolution. The best way to counter this is by incorporating two joints in the shaft assembly. If the design can be set up so both joints operate at the same angle, the lead/lag function can be removed from the output of the set by proper phasing. (*Figure 30*) This leaves the center portion of the shaft the only part running at a non-constant velocity. Torsional vibrations are thus reduced to the forces required to accelerate and decelerate the center portion of the shaft. Reducing the torsional moment of inertia of the center portion of the shaft will obviously reduce this forcing function as will lower joint set angles.

If both joint angles are in the same plane, they can achieve cancellation as long as they are equal, i.e. one up, one down, or both down, both up.

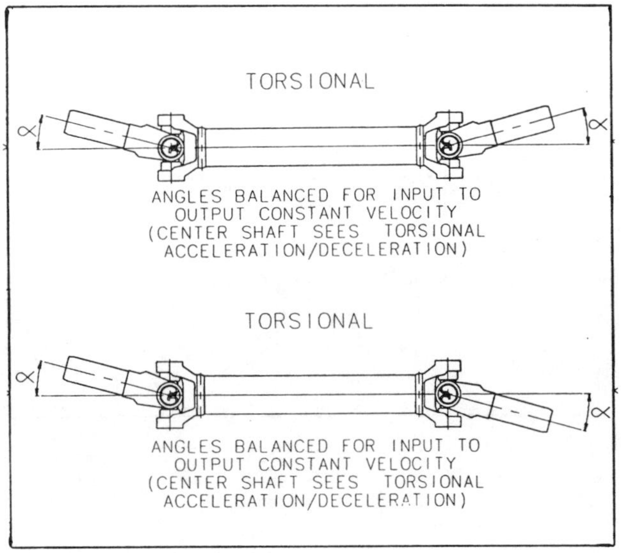

Figure 30
Joint Angles in the Same Plane
(e.g. the plane of the paper)

To set up a propeller shaft system that does not have lead/lag between the input and output shafts, you must phase the yokes. Phasing is the angular adjustment of the yokes on an intermediate shaft to obtain constant velocity between the input and output shafts.

Visualize a three dimensional system. (*Figure 31*) Let the center portion of a two joint shaft be horizontal. If we look at this center portion from the output end we note:

1. The input yoke center line angles upward and to the left.
2. The output yoke center line angles upward and to the right.
3. The plane formed by the input yoke and the center portion tilts left of vertical.
4. The plane formed by the output yoke and the center portion tilts to the right of vertical.

The angle between these two planes is known as the ***phase angle***, Now we will mount the two yokes of the center section as follows:

a. The input yoke's ears are parallel to the plane of the input joint angle.

b. The output yoke's ears are parallel to the plane of the output angle.

c. The angle between these two yokes is also the ***phase angle.***

If the true joint angle of the input joint is equal to the true joint angle of the output joint, and the center shaft yokes are set at the phase angle, the assembly will deliver constant velocity from the input to the output. The intermediate shaft, however, will lead and lag every 180° as dictated by the true joint angles.

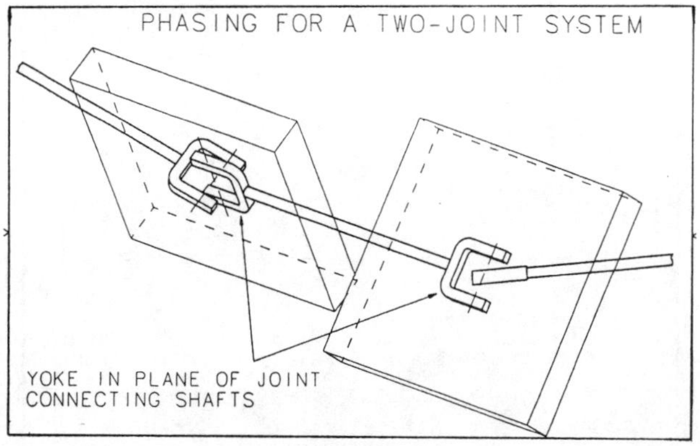

Figure 31
Phasing Visualization

(b) **Secondary Couple** — The secondary couple is the moment created by opposite trunnions of the cross when the plane of all four trunnions is not perpendicular to the centerline of that portion of the shaft.

When the joint angle is zero there is no secondary couple, i.e. it is zero. As the joint angle approaches 90° all torque goes into the secondary couple and there is no primary couple or turning force.

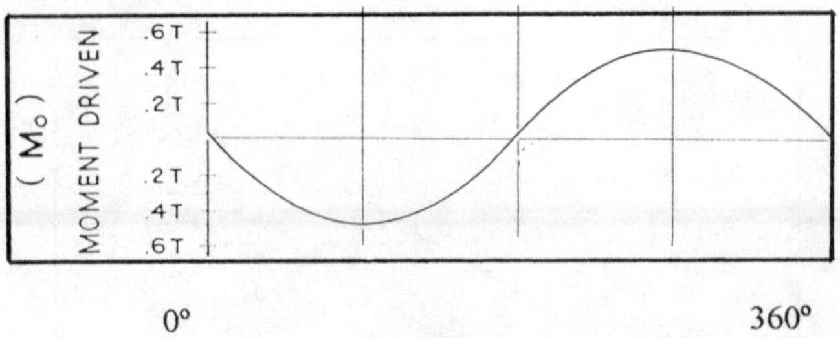

Figure 32
Moment Produced in Driven Member by Secondary Couple

From the above we can see there is a change in transmitted torque associated with the lead/lag phenomenon. This is, of course, a part of the vibration forcing function.

The secondary couple exists as rotating vectors at the bearing, acting in a circle tangent to the center of the bearing, with the center of the circle on a line perpendicular to the plane of the true joint angle. (*Figures 33, 34 and 35*)

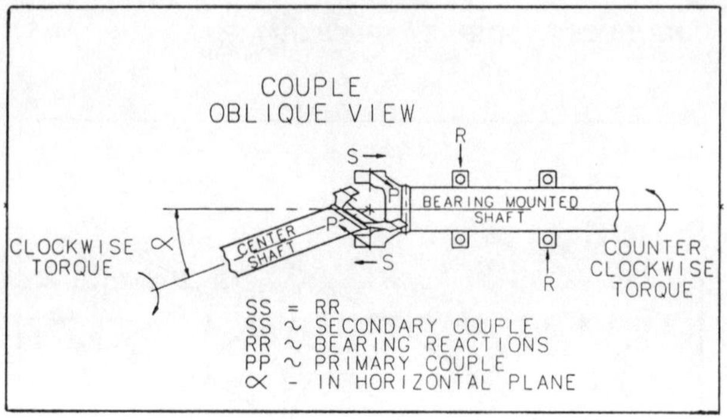

Figure 33
Oblique View Showing Generation of Secondary Couple

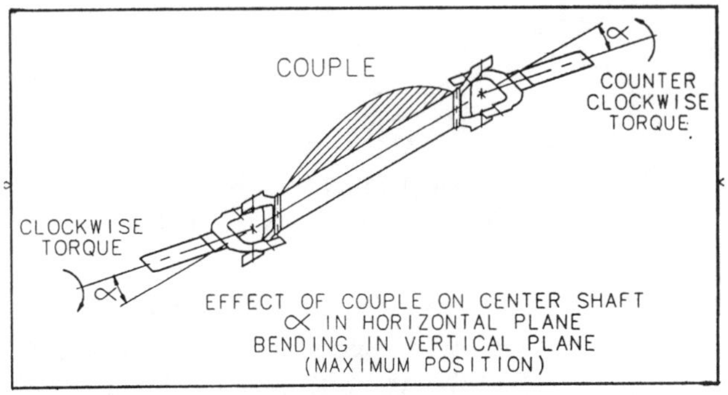

Figure 34
Bending Effect on Center Shaft with Equal Angle Joints in Opposite Direction

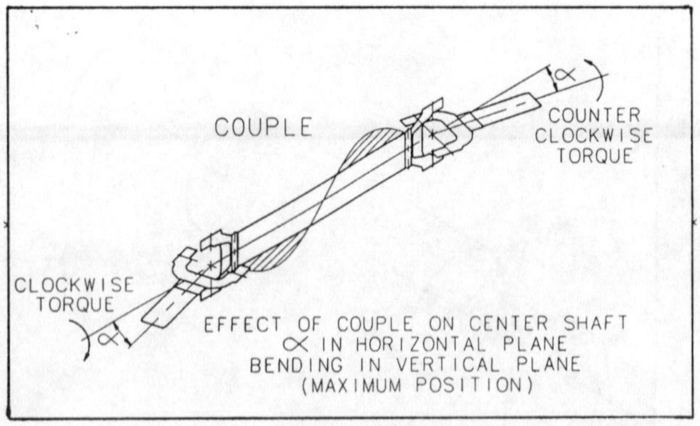

Figure 35
Bending Effect on Center Shaft with Equal Angle Joints in Same Direction

Out Of Balance — The amount the shaft is out-of-balance will produce rotating vectors at each end. Since a shaft is long relative to its diameter, dynamic balancing is imperative. The rotating vector at each end can be different and out-of-phase. Corrections to normal out-of-balance due to tolerances, tube irregularities, and yoke irregularities can be made by adding balance weights at either end as needed to dynamically balance the shaft. Normal industry specs may be as little as 1.77 to 3.53 N-mm. The tighter the spec, the more the cost.

Static and dynamic balance are not the same. A three dimensional rotating body can be in static balance and not be in dynamic balance. To illustrate, consider a cylindrical drum mounted on a shaft with bearings. (*Figure 36*)

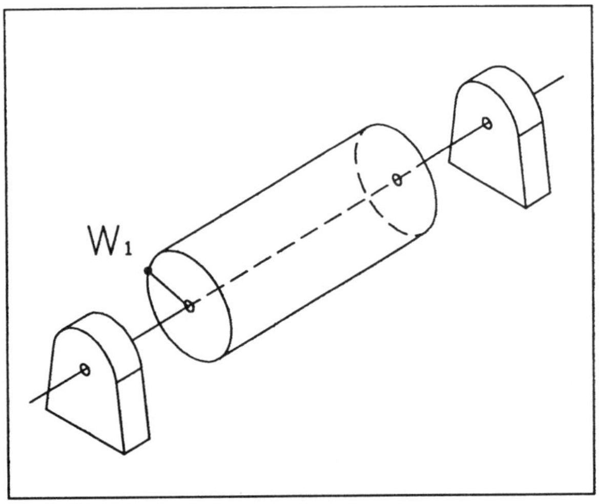

Figure 36
Drum Out-of-Balance

It has a concentrated mass at one end near the rim. If the bearings turn freely, the drum will turn and the concentrated weight will eventually come to rest at the bottom. (*Figure 37*)

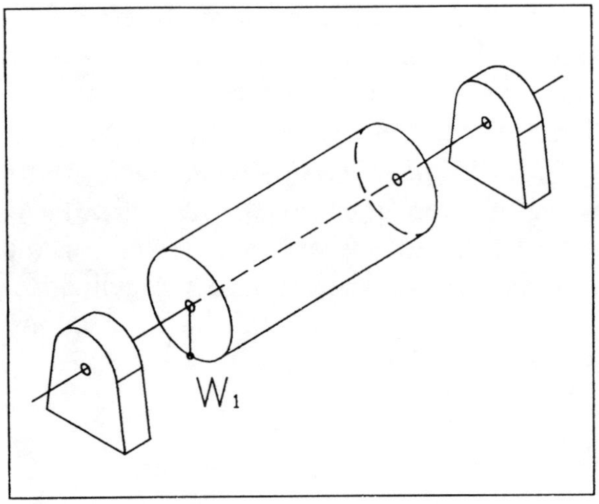

Figure 37
Drum Always Comes to Rest with Weight at Bottom

Now if we add an equal weight diametrically opposite but at the other end, the system will be in ***static*** balance and the drum can come to rest in any position. (*Figure 38*)

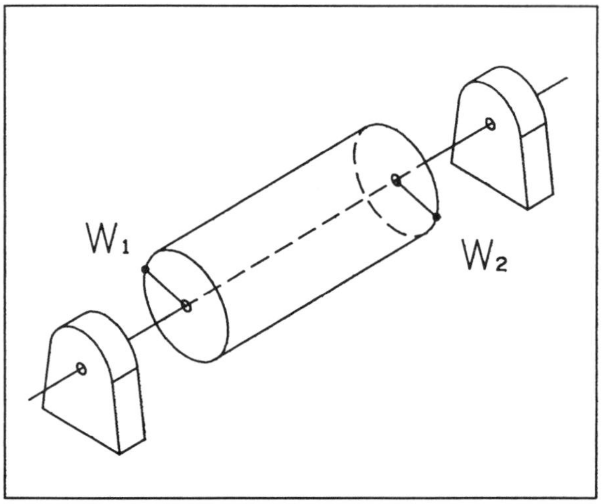

Figure 38
Drum in Static Balance
(Can Come to Rest in Any Position)

Spinning the drum, however, will produce centrifugal vectors F_1 and F_2 directly opposite W_1 and W_2. The system is not in ***dynamic*** balance and will vibrate as it continues to spin. (*Figure 39*)

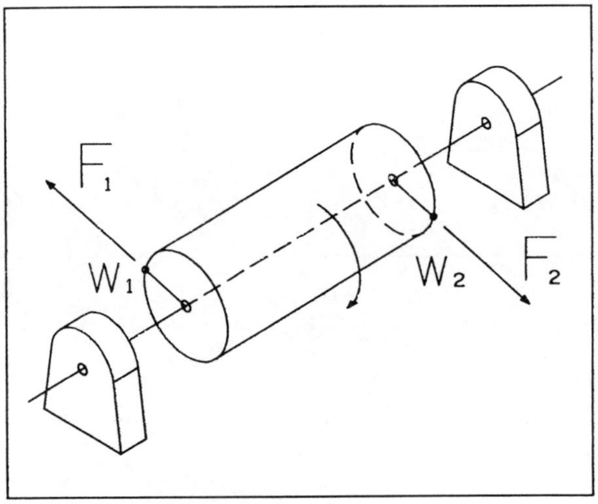

Figure 39
Rotating Drum Showing
Centrifugal Force Vectors F_1 and F_2

If W_1 and W_2 are a permanent part of the drum, we will need to add equivalent weights, $W_{1\text{-}1}$ and $W_{2\text{-}2}$ directly opposite W_1 and W_2 to achieve dynamic balance. The system now will come to rest in any position (static balance) and can be rotated without vibrating (dynamic balance). (*Figure 40*)

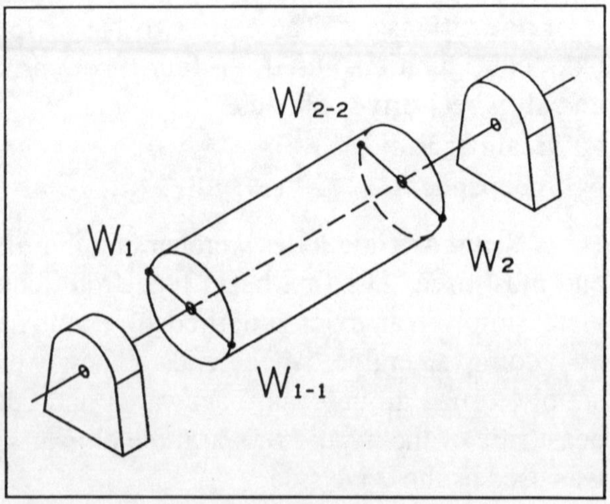

Figure 40
Drum Corrected for Both Static and Dynamic Balance

NOTE: Many tire balancing machines are only spin balancers. They use the spinning to generate centrifugal force and enhance the out-of-balance vector so as to better place the weight for static balance. Since they sense only in one plane, they do not address true dynamic balance and do not indicate the opposing force vectors associated with dynamic balance.

FRONT AXLE — As new front axle designs are produced gear sets are designed specifically for front operation. Many front axles are still using gear sets designed for rear axle operation thus running on the coast side of the gears rather than the drive side.

Limited slip devices are generally not employed in front axles. The reason is that the restriction of differentiation can cause adverse handling issues. The turning radius is affected and changed when the device becomes active. Recently there have been several designs introduced that are configured so they will not restrict differentiation within the full steering range of the vehicle. In this case they will only become active if there is an actual spin out of a wheel that exceeds the differential action required at full lock steer.

Configurations — There are two basic front axle configurations used in four-wheel-drive and all-wheel-drive vehicles.

- Beam or Rigid
- Independent

Beam or rigid axles were among the first and most used. The first beam type front axles were simply rear axles modified for front use by adding steerable wheel ends. They were simply turned around, run backwards and the coast side of the hypoid or spiral bevel gear set was used as the drive side.

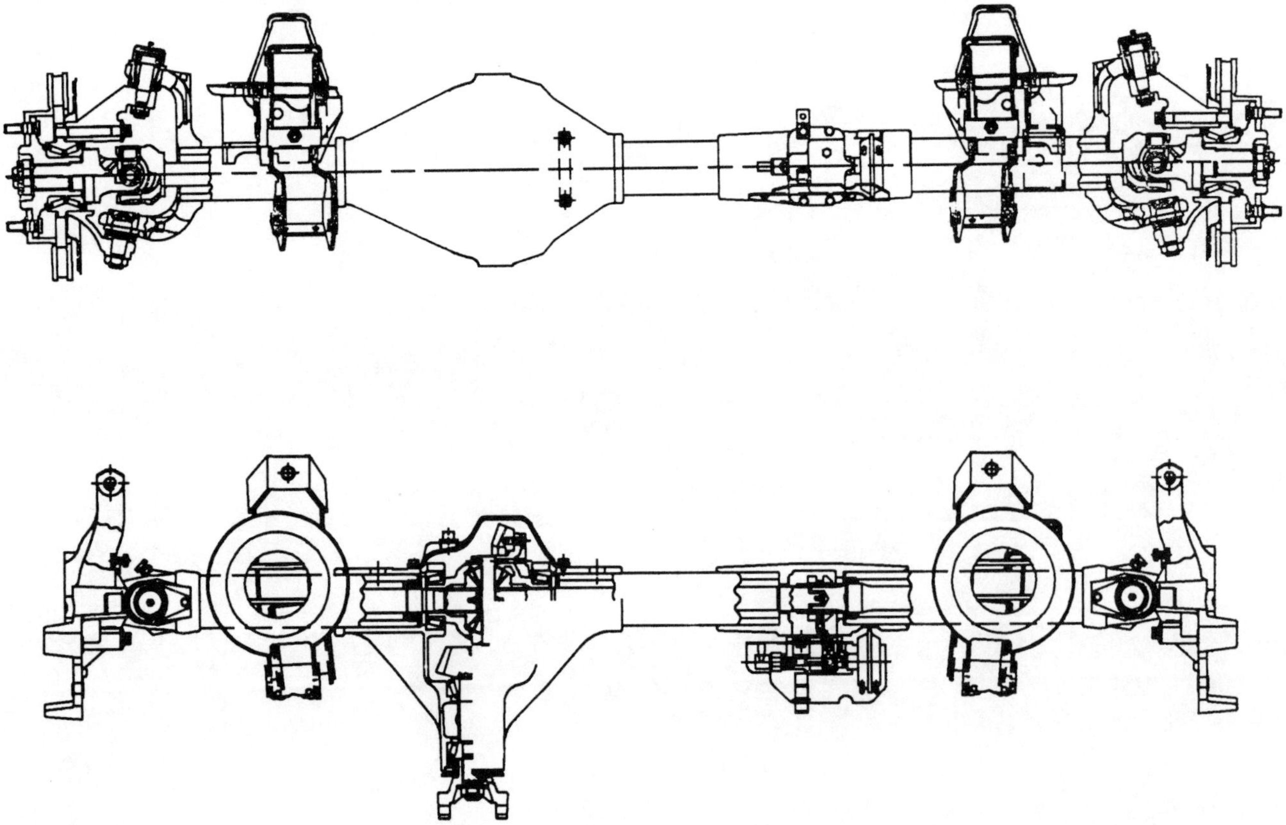

Figure 41
Two Views of a Beam or Rigid Front Axle

Independent wheel suspension front axles were created by frame mounting the center section of the axle. The sides were cropped off short and a flange was provided so a short shaft known as a half shaft could be attached to drive the independently sprung steerable wheel end.

Some front axle designs have the center section attached to the engine, with one of the axle shafts actually running through the engine's oil pan. An advantage of this design is that the front of the vehicle can be substantially lowered.

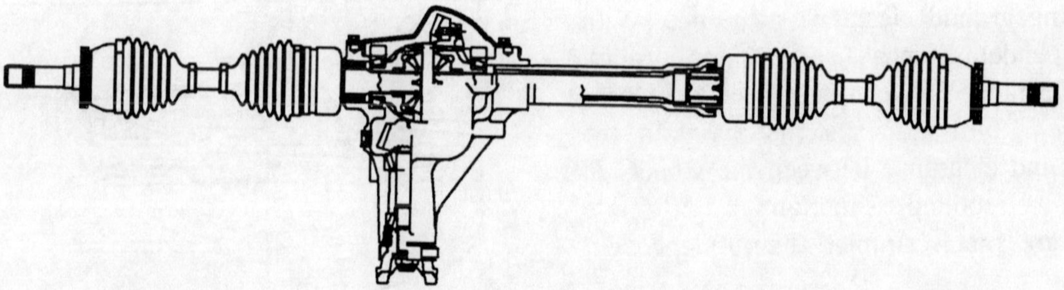

Figure 42
Independent Front Axle

The **twin beam** configuration is a special case of an independent front axle system. It has the carrier member mounted in one side of the twin beam and a joint in the center, so the opposite beam can be supplied with power.

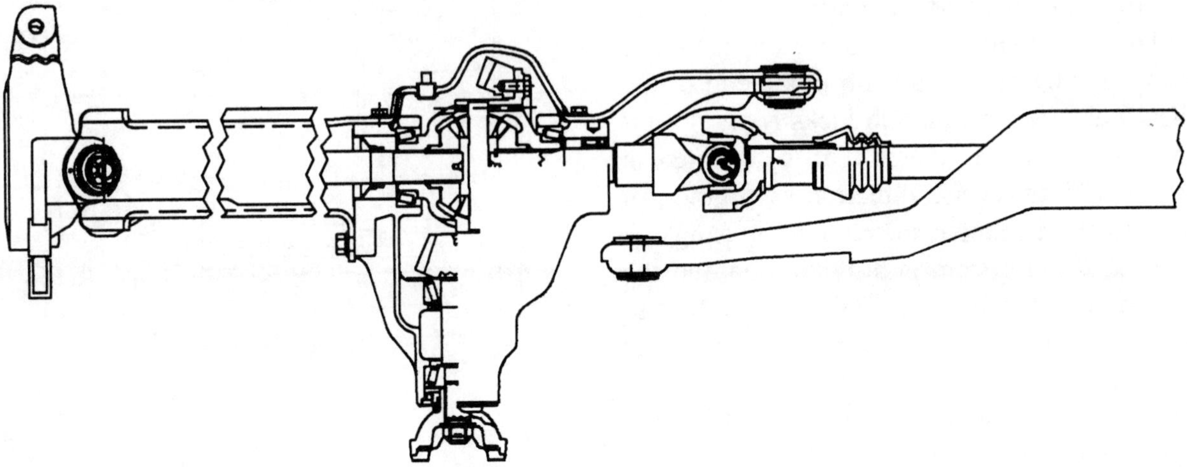

Figure 43
Twin Beam Front Axle

Advantages & Disadvantages of Rigid vs Independent Axles — Axle types that allow independent suspension are generally thought to provide a better vehicle ride because there is much less unsprung weight. They also allow a wider variety of suspension types, giving the designer of the vehicle many options. Ride, handling, and packaging of other components can be approached in an integrated manner.

Independents are at a disadvantage when considering ground clearance in jounce. With an independent wheel suspension system, ground clearance between the wheels decreases in jounce. Rigid axle systems maintain the same ground clearance between the wheels in any jounce/rebound situation. Rigid axle systems are much simpler than independent ones.

Disconnects — Disconnect devices are often used in front axles. Their purpose is to stop rotation of the front drive system, thus reducing noise and vibration forcing functions. Since the front driving system is not running at highway speeds, balance of components and joint phasing are not critical.

Typically there are two types of disconnects in use: (1) hub locks that disconnect the driving hub right at the wheel, and (2) clutching devices that disengage one axle shaft.

Both manual and automatic hub locks are available. Manual hub locks are engaged and disengaged by turning a knob in the end of the wheel hub. Automatic hub locks contain a bi-directional one way clutch that is activated in the forward or reverse direction by rotation of the axle shaft relative to the axle housing. If the front drive system is disconnected and the axle shaft is not turning, the wheel is free to turn; if the axle shaft is driven from the engine, the clutch engages to drive the wheel.

Some hub lock systems have also been designed to be controlled by the driver from his/her position behind the steering wheel. Vacuum, air, hydraulic, and electric actuator systems can be employed for this purpose.

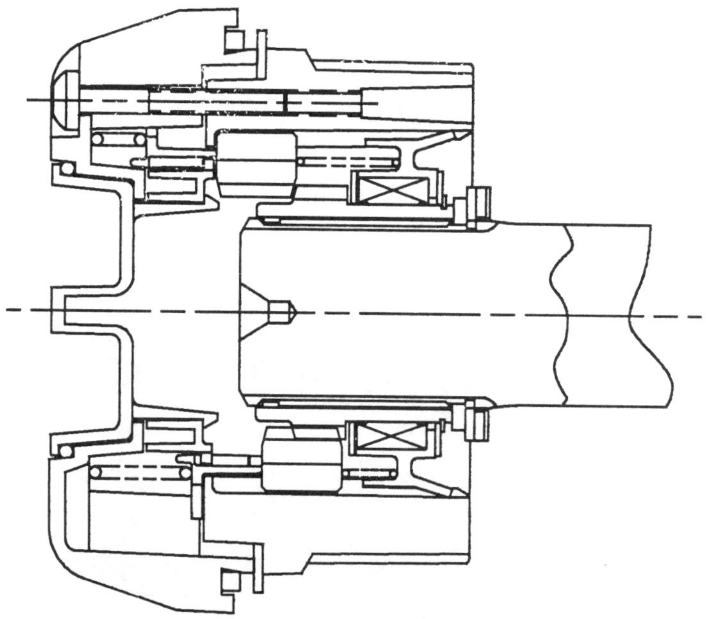

Figure 44
Manual Locking Hub Lock

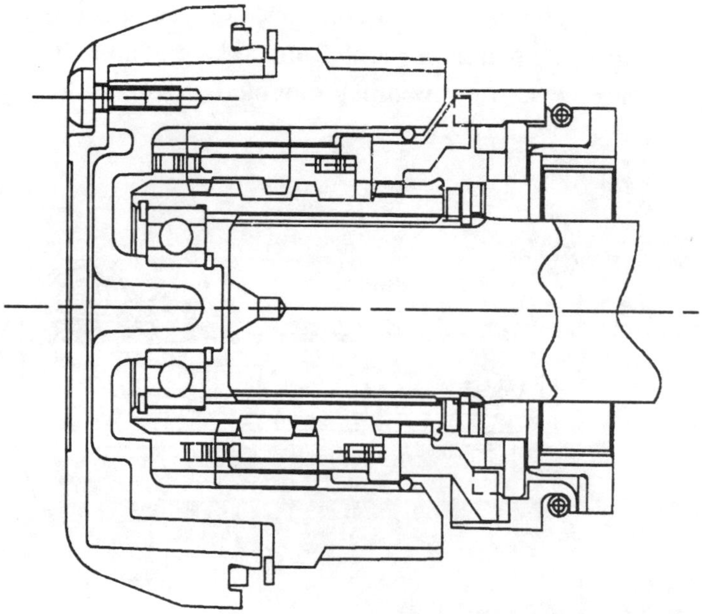

Figure 45
Semi-Automatic Locking Hub Lock

The second method of disengaging the front wheels is to put a clutching device in one or both of the front axle shafts. This is usually a sliding tooth collar clutch, moved by a shift fork. Actuation is by vacuum, electric, or other means and is normally driver controlled by a switch or lever.

If a single side disconnect is used the differential gears will normally rotate, while the ring gear and pinion remain stationary. This reduces stirring action of the axle oil. Seal and bearing drag of the non-rotating components is zero. The continuous rotation of the differential gears can produce some objectionable noise. To reduce this noise it may be necessary to use higher quality gears than are normally used in differential gear sets. In my opinion it would be better to spend money on two disconnects than to spend it on improving differential gear quality (assuming the manufacturer really wants to carry the weight of a front axle, propshaft, transfer case, and disconnect and not use them).

Even better would be to trade the cost and weight of a hublock or disconnect system for an inter-axle differential in the transfer case, making the system full-time four-wheel-drive. Further reduction of weight and cost would be possible. Also, most full time systems would not be required to sustain the wind-up loads in turning as part-time-systems do. The driveline could be downsized to meet the available engine power and wheel traction. In short the full time system offers more value and less cost.

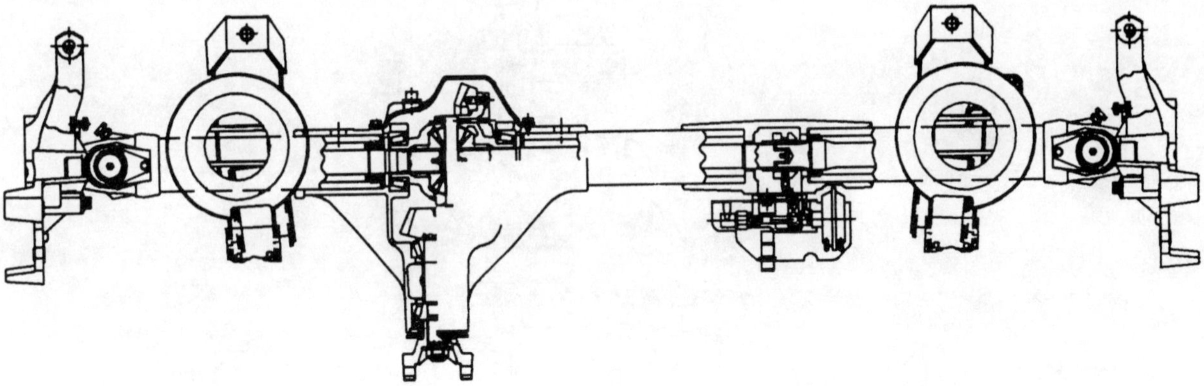

Figure 46
Axle with Clutch Type Disconnect Device

REAR AXLE — As in front axles, both beam and independent type rear axles are being used. Independent suspensions are becoming more popular for rears in sport utility vehicles due to the perceived advantages in terms of ride and handling. Beam type rear axles are still popular in pickup trucks and in some sport utility vehicles. . The simplicity and robustness of beam type axles will continue to make them the popular choice for trucks, especially in the larger sizes where they see a greater amount of commercial use.

Axle ratios normally are kept the same for both front and rear axles. In part-time or locked mode this will keep wind up of the drivetrain to a minimum. In full-time systems (those with differentials), the amount of differential action will be minimized.

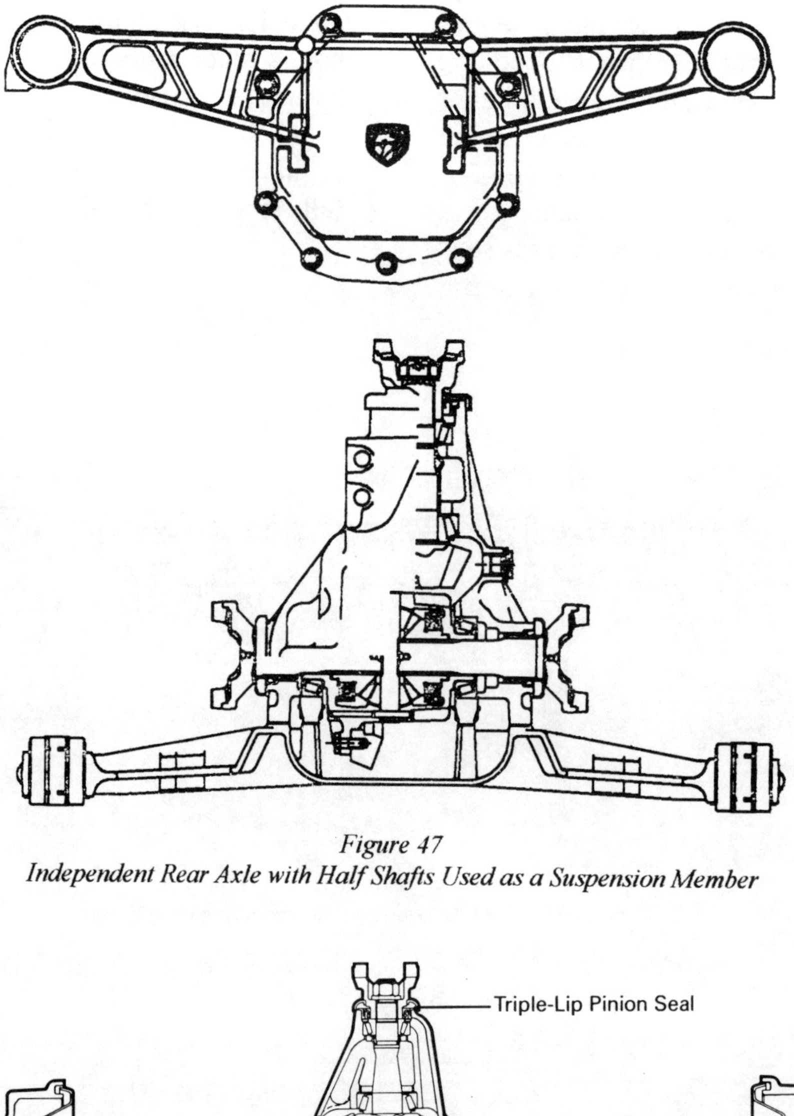

Figure 47
Independent Rear Axle with Half Shafts Used as a Suspension Member

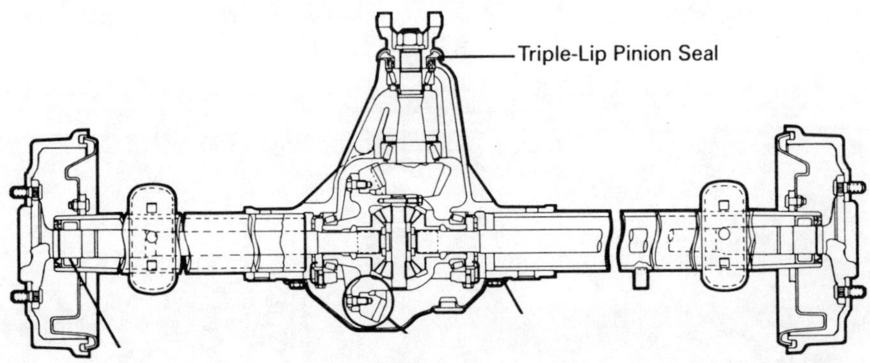

Figure 48
Beam Type Rear Axle

HALF SHAFTS — The shaft that runs between a rigid mounted axle center section and an independent wheel end is known as a half shaft. These shafts may have cardan joints at both ends, both CV joints or a combination of CV and cardan joints.

On a steering axle it is desirable to have CV joints at the outer end to avoid the lumpy feel when cornering that is associated with the lead/lag phenomenon of cardan joints. In commercial vehicles that are used for more rugged applications this may not be necessary.

On fronts there still may be an issue if cardans are used inboard with CVs outboard. The inboard joint angles due to jounce, rebound, or vehicle weight variation due to loading can produce noticeable and undesirable torque fluctuations.

For rear half shafts and in the inboard location of front shafts, cardan type joints may be used. On rears the joint angles can be properly designed and phased so as to eliminate most problems caused by the non-CV characteristics of the cardan joint.

Plunging type ball joints or tripod type are used where the suspension requires variable shaft length due to steering or jounce and rebound. There have been some suspension systems designed using the half shaft as a tension/compression suspension member. In this case a non-slipping type joint is of course employed.

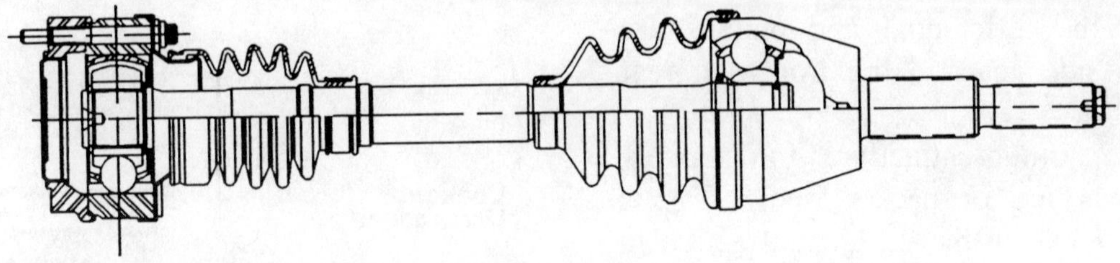

Figure 49
Half Shaft with Plunging CV at Axle End and Rigid CV at Wheel End

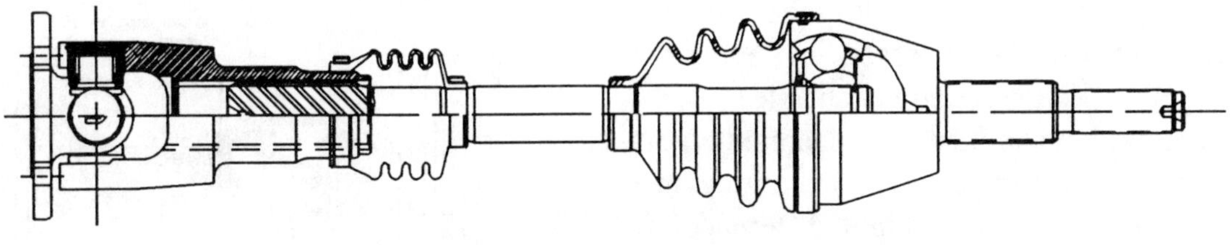

Figure 50
Half Shaft with Cardan Inner Joint and CV at Wheel End

TANDEM AXLE AS A FOUR-WHEEL-DRIVE — A tandem axle system is the dual axle found under the rear of most Class 8 semi-tractors. Each axle has two tires at each end so there is a total of eight wheels. Pairs of tires act as a single wheel, so in effect it is really a four-wheel-drive system.

Some early tandem axles drove only the front axle and in some cases a heavy duty belt drive was provided to drive the rear axle from the front axle. The belt was usually positioned between the dual tires. With a belt system there was no differential action provided between the front and rear axle.

A better performing system has a differential incorporated into the front axle of the tandem axle pair. This is usually a bevel gear type differential and provides a 50/50 torque split between the two axles. Torque from the propeller shaft drives the carrier member of the differential and outputs are through the side gears to the front and rear axle.

For maximum traction in slippery conditions, as ice or packed snow, it is desirable to lock out the differential action in order to maximize the drawbar pulling capability of the system. Typically the system is locked by the use of a sliding collar that connects the input to one of the side gears. This will force both axle pinions to rotate at the same speed. Each axle also has a bevel gear differential so side to side differentiation can occur in either axle. Further traction enhancement can be obtained by adding a differential locking device in each axle.

The system should not be operated on dry pavement with the differential locking devices engaged. If this is done slight differences in tire size can cause a wind up of the drivetrain system. This windup or trapped load will be added to the torque required to propel the vehicle thus causing all gears, shafts etc. to run at a higher load than necessary.

Useful life of the system can be shortened due to running in locked mode when it is not necessary to provide maximum traction. For this reason the vehicle manufacturer usually recommends to use the locked mode only when there is ice, snow, or other extremely slippery conditions.

Figure 51-a shows a typical tandem axle setup. Note the inter-axle differential at the input of the front axle of the tandem pair. This differential has the locking collar moved by a shift fork that is usually air actuated. In this unit each axle has a planetary gear reduction in addition to the bevel gear differential. This is a shiftable unit that can be used or bypassed, giving the vehicle an additional set of overall ratios. It can be used for starting, for low speed operation as in backing and docking, or it can be used as a splitter or range for the primary transmission.

Figure 51-a also shows the torque flow with the inter-axle differential in operation. Torque will be balanced, front to rear.

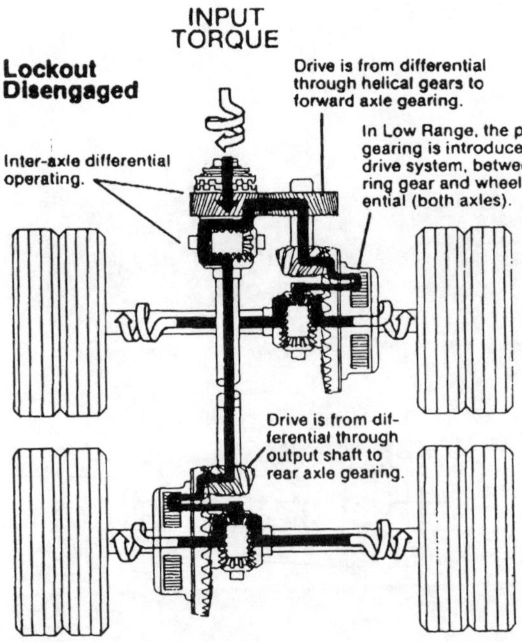

Torque is transmitted to both axles through inter-axle differential action.

Figure 51-a
Tandem Axle with Differential Operating

Figure 51-b shows the inter-axle differential locked. Torque to the front and rear axles will be dependent on the coefficient of friction under the tires and the tire size.

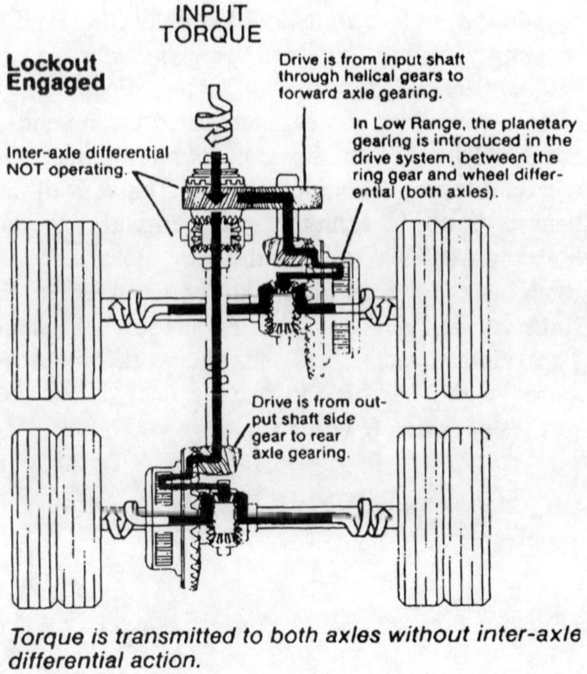

Figure 51-b
Tandem Axle with Differential Locked

Figure 52 is a cross section of the front axle of a tandem pair, showing detail of the inter-axle differential and locking collar.

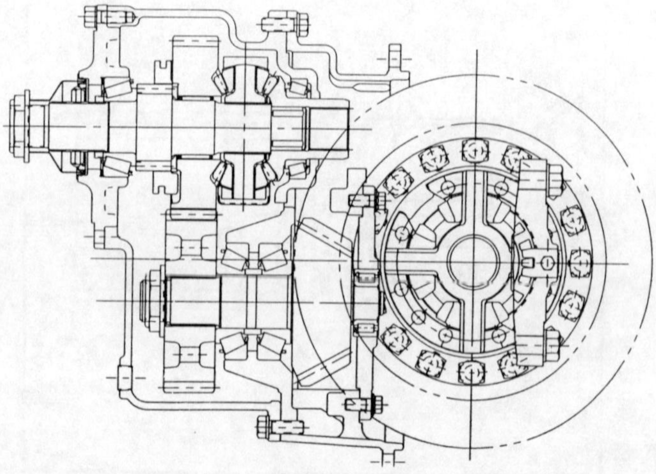

Figure 52
Front Axle of Tandem Pair

DIFFERENTIAL

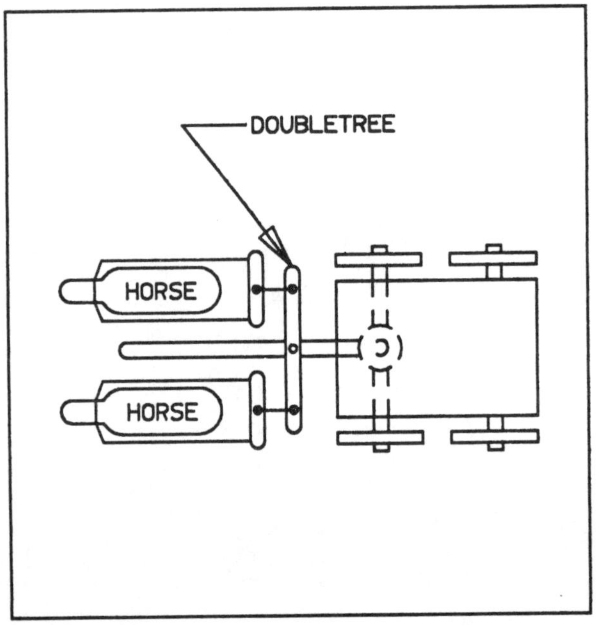

Figure 53

The following is an explanation I have found useful in describing the function and need for differentials.

The differential is not a modern device. It had its origin in the horse drawn wagon. (*Figure 54*) To divide the work equally between both horses of a team, a double tree is employed. It is simply a bar mounted perpendicular to the wagon's tongue, but free to pivot. Each horse's collar and pulling harness is attached to one end of the doubletree. If one horse is not pulling his share, the doubletree will pivot and that horse will fall behind. Since he wants to stay even with the other horse he will have to expend a little more energy to catch back up. Note, during the above scenario, the force vector produced by each horse stays the same. The velocity vector, and thus the work or energy, varies slightly as one horse falls behind and then catches back up. (*Figure 53*)

Seattle Post-Intelligencer Library

Figure 54
Typical Harness Arrangment. (Note doubletree behind horse team.)

PURPOSE OF A DIFFERENTIAL – Why do we need a differential and how does it work? When a vehicle is turning a corner the outside wheel is at a larger radius than the inside wheel. (*See Figure 55*) That is, the wheel farther from the center of the turning radius will be going around a larger circle, moving faster, and turning a greater number of revolutions than the inside wheel. If both wheels were connected to the same axle shaft there would then be a skidding or slipping action required of one of or both wheels in order to negotiate the corner. To apply torque to both wheels, a torque dividing device is required. This device is the differential.

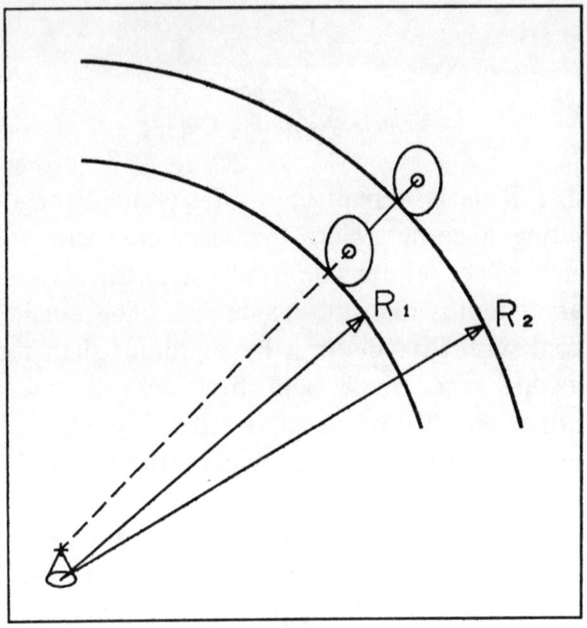

Figure 55
Axle Going Around Corner

A differential functions exactly like a simple balance scale. The force at the fulcrum is exactly equal to the force in the pans. Since the pans are equidistant from the fulcrum, they will each support one half the force at the fulcrum. (*Figure 56*)

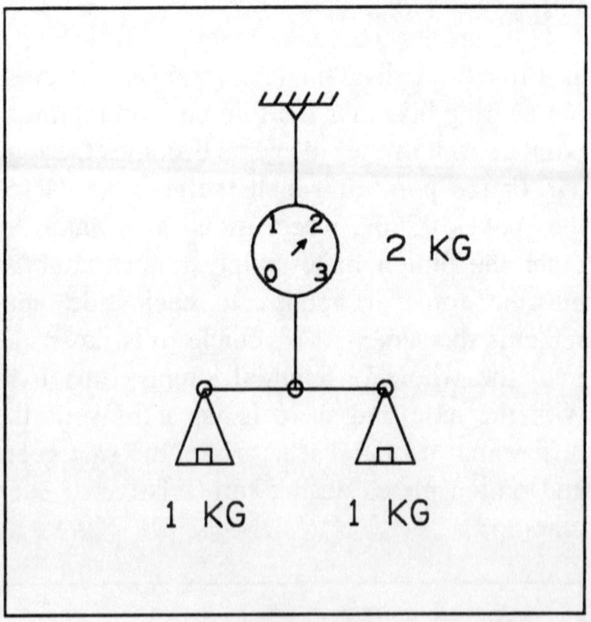

Figure 56
Balance Beam

Assume we have two spoked wheels mounted on an axle so they can turn freely relative to each other. We insert a stick in the spokes near the rim and parallel to the axle at a point farthest above the ground. We tie a cord to the center of the stick and pull in the direction we want the vehicle to move. An equal driving force or torque will be applied to each wheel. (*Figure 57*)

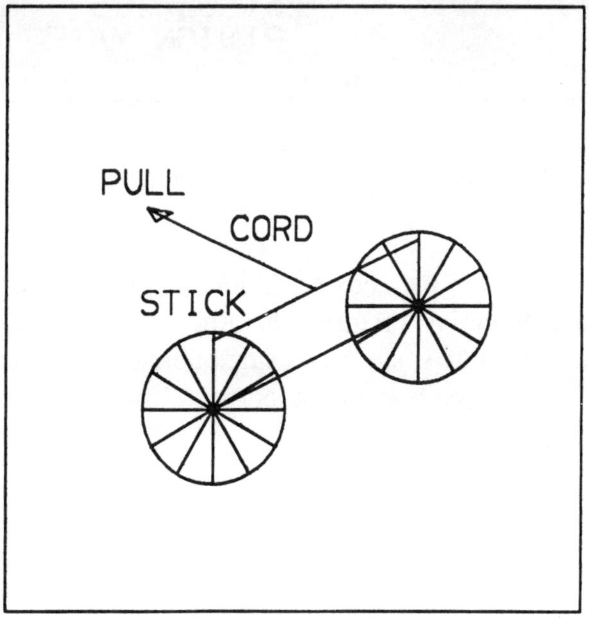

Figure 57
Axle with Wheels Receiving Equal Torque

Consider the pinion mate of a differential as a free body diagram. (*Figure 58*) The cross pin bearing becomes the fulcrum and the mesh point at each of the side gears becomes the load pan or the point at which the stick contacted the spokes. If one wheel turns faster than the other the pinion mate simply rotates slightly, but the force it applies to each side gear remains the same. As a vehicle rolls down the road the whole differential simply rolls over with the axle and there is no activity in the differential itself. That is to say, the side gears and pinion mates do not turn relative to each other.

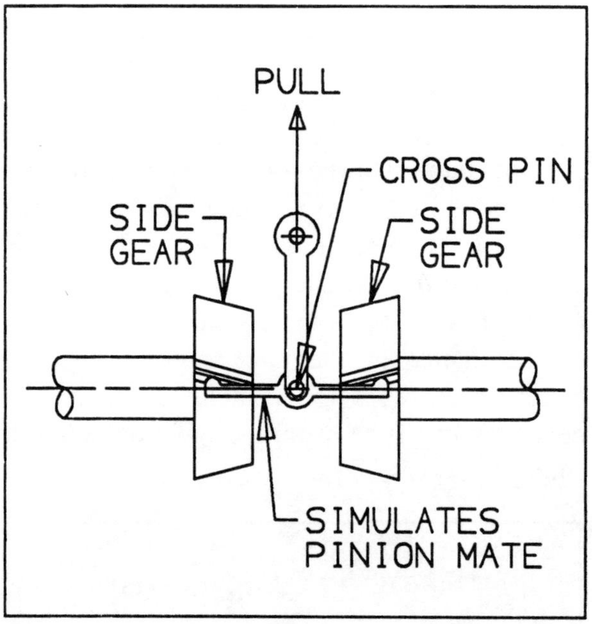

Figure 58
Differential with Pinion Mate Replaced by Balance Beam

From the above it should be obvious that a differential is required in either a front or rear driving axle since the outside of either must run faster when turning a corner. Now look at *Figure 59*.

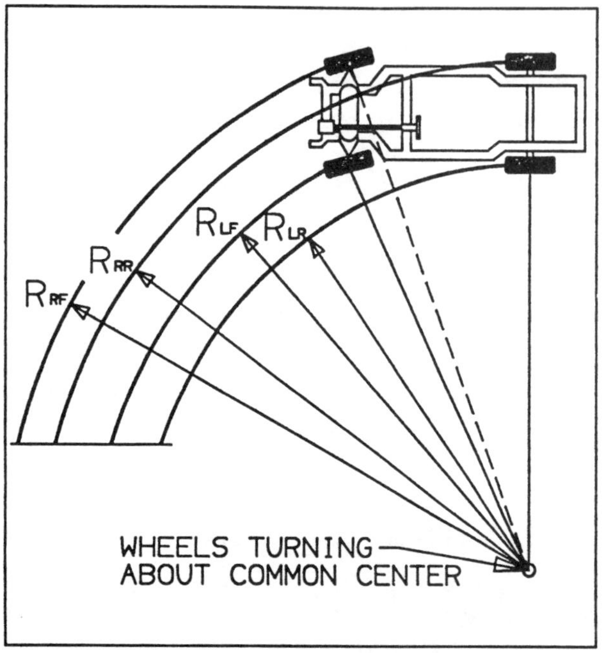

Figure 59
Vehicle Negotiating Curve

This is a plan view of a vehicle negotiating a corner. Note the respective radii of each wheel. The inside front wheel travels in a larger radius than the inside rear. The outside front wheel travels in a larger radius than the outside rear. Since both front wheels travel farther than their respective rears, it is obvious the front of the vehicle travels farther than the rear.

If this is a four-wheel-drive vehicle and we have equal axle ratios and nearly equal tire rolling radii, the front propshaft will need to rotate faster than the rear propshaft while negotiating that corner. That's why we need an inter-axle differential. However, when the vehicle is going straight both propshafts will be turning at nearly the same speed.

Figure 60-a is a balance beam analogy of a full-time four-wheel-drive system with three differentials: one for each axle and one in the transfer case. The center fulcrum represents the differential in the transfer case, and the front and rear fulcrums represent the axle differentials.

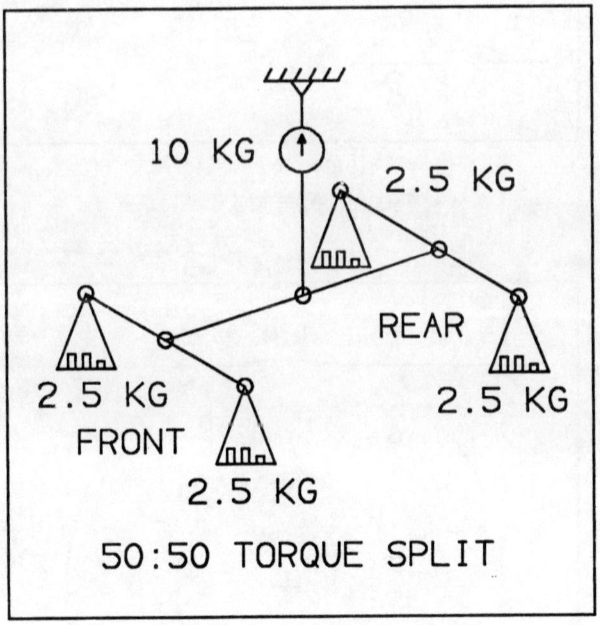

Figure 60-a
Balance Beam Analogy of Full-time Four-Wheel-Drive System (with 3 differentials or fulcrums)

To tailor a vehicle for specific handling or mobility requirements, it may be desirable to have a center differential (transfer case differential) that will provide an unequal amount of torque to the front and rear axles. This can be done by using an epicyclic or planetary gear set for the differential in place of the traditional bevel gear set type of differential. The effect is the same as moving the fulcrum forward or aft in the balance beam analogy. (*Figures 60-b and 60-c*)

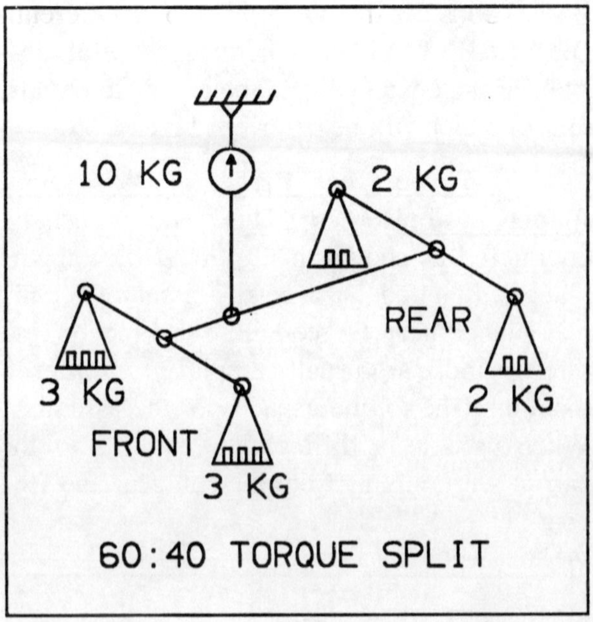

Figure 60–b
Fulcrum Moved Forward to Simulate 60/40 Torque Split, Front to Rear

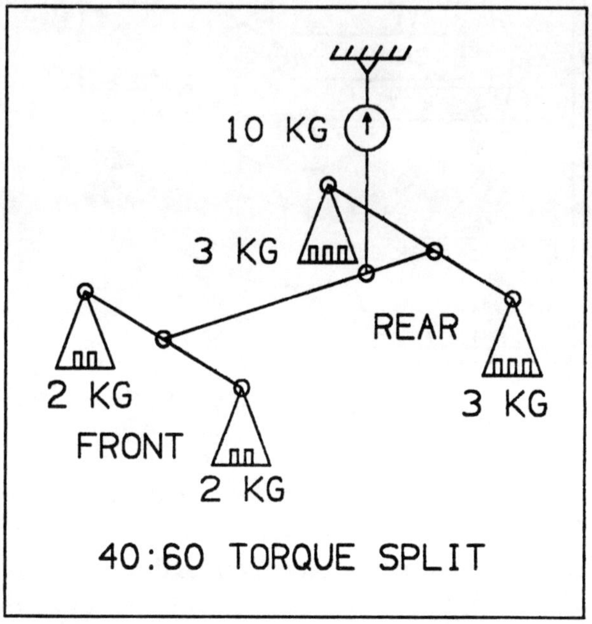

Figure 60–c
Fulcrum Moved Aft to Simulate 40/60 Torque Split, Front to Rear

TYPES OF PLANETARY DIFFERENTIALS – There are several possible arrangements for planetary differentials, four of which will be described here:

1. **<u>Three Gear Planetary With Sun, Planets and Ring</u>** – This type is usually configured as shown in *Figure 61-a*, with the planets mounted on a carrier assembly. Only one planet gear is required for function but three or more are usually employed in order to maximize the torque capacity of the assembly. When used as a differential, input is to the carrier with outputs from the sun gear and ring gear.

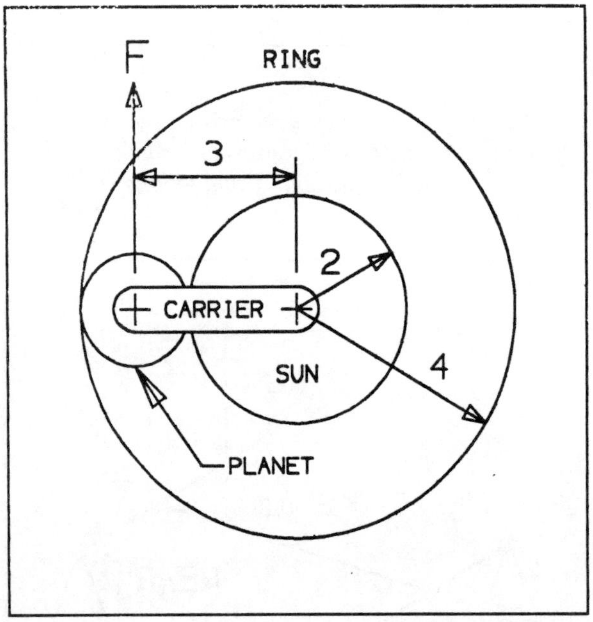

Figure 61-a
3 Gear Planetary with Sun, Planets, and Ring

Looking at a free body diagram of the planet (*Figure 61-b*) we have an input load vector acting on the planet pin and reaction (or output) vectors where the planet meshes with the sun and ring. The output vectors are equal and are each one half of the input vector. Since the applied force to the ring gear (*Figure 61-c*) acts at a larger radius than the one acting on the sun, the ring will be at higher torque than the sun. Ring gear torque will be ring gear pitch radius times F/2 and sun gear torque will be sun gear pitch radius times F/2.

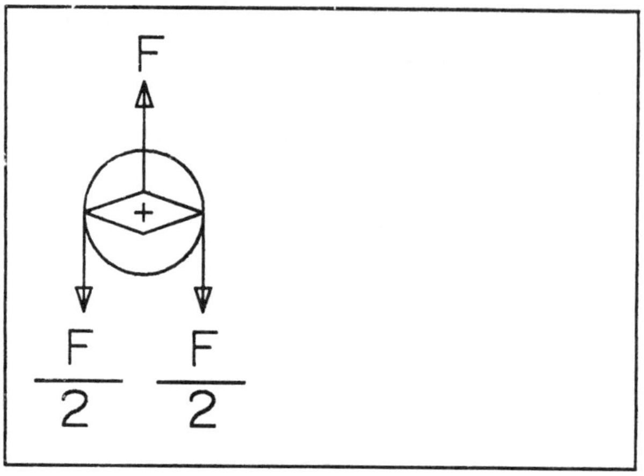

Figure 61-b
Free Body Diagram of Planet Gear

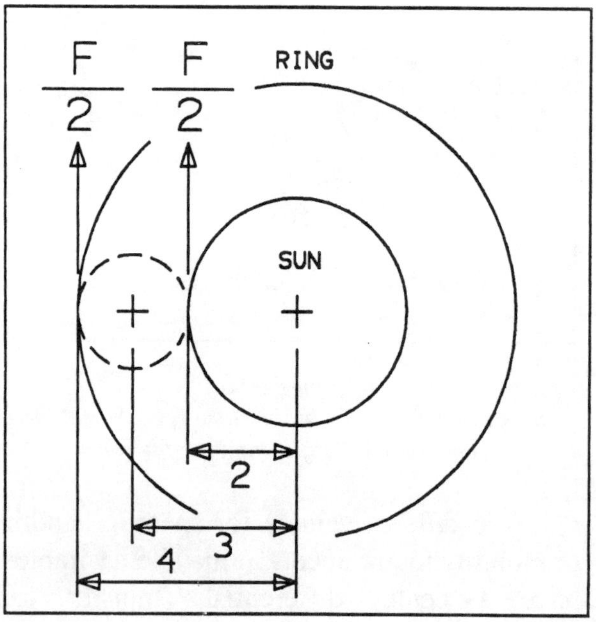

Figure 61-c
Amount of Torque Applied to Ring and Sun

<u>Example</u>

Let: Planet Gear radius = 1
Sun Gear radius = 2
Ring Gear radius = 4
Tangential Force = F

Then: Input Torque = 3 F (Carrier Torque)

Ring Torque = $\frac{F}{2} * 4 = 2F$ (2/3 Carrier Torque)

Sun Torque = $\frac{F}{2} * 2 = F$ (1/3 Carrier Torque)

2. **Four Gear Planetary With Sun, Two Intermeshing Planets and One Ring Gear** — This type requires the ring gear to be the input with the sun and carrier acting as the two outputs. Ring gear torque will equal carrier torque plus sun torque. Sun torque will be F1 times Rs and planet torque will be ring gear torque minus sun gear torque. (*Figure 62*)

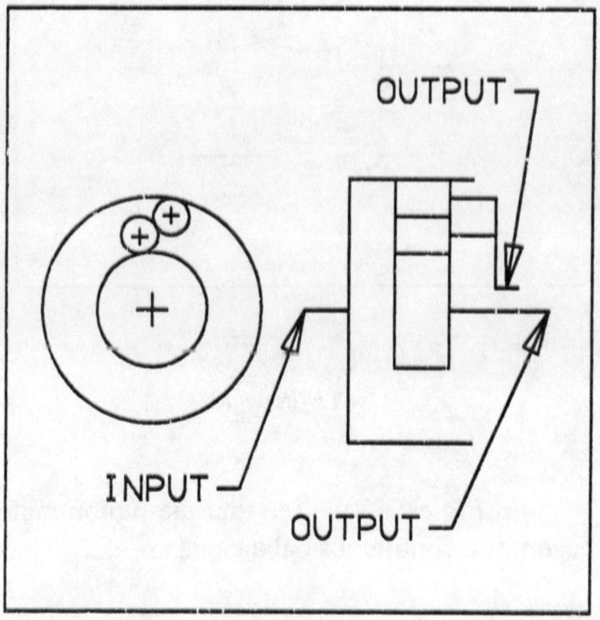

Figure 62
4 Gear Planetary Differential with Ring Gear Input

3. **Four Gear Planetary With Two Suns and Two Planets** — These may have multiple sets of planets for load sharing to reduce gear stress. They are set up with equal diameter sun gears in order to give an equal torque split like a typical bevel gear differential. It is possible however to make one sun larger and in this case the ratio of torques will be directly proportional to the ratio of the suns' diameters. (*Figure 63*)

4. **Four Gear Planetary With Two Ring Gears and Two Intermeshing Planets** — This is an inversion of the system No. 3. There are two rings instead of two planets. Torque is a function of the ring gear diameters and directly proportional to the diameters. (*Figure 64*)

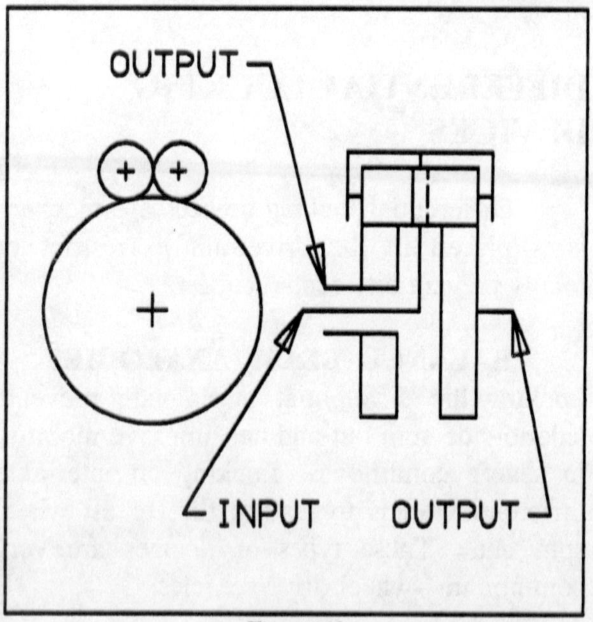

Figure 63
4 Gear Planetary Differential with Dual Sun Output

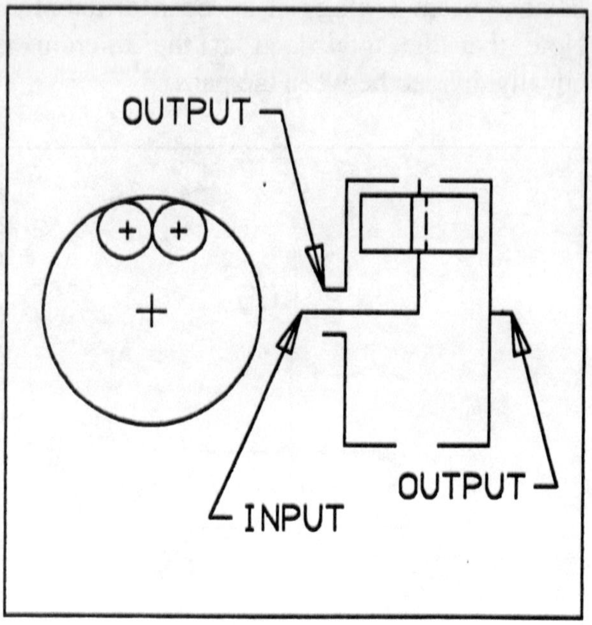

Figure 64
4 Gear Planetary Differential with Dual Ring Gear Outputs

In all of the above gear sets multiple sets of planet gears can be used in order to increase torque capacity. Single planet sets are shown for simplicity of analysis. The type of differential selected will be a composite of space and load carrying capacity required. Other unique arrangements of gears are of course possible and may have merit in any particular situation.

DIFFERENTIAL LOCKING DEVICES

Differential locking devices are mechanisms placed in the drivetrain to restrict or totally prevent differential action.

BALANCE BEAM ANALOGIES — Locking the differential of an axle prevents side-to-side spin out and can improve mobility in severe conditions. Locking an inter-axle differential tends to restrict fore to aft wheel spin out. These types of devices are very common in 4-wheel-drive vehicles.

Open Differential — *Figure 65* is a balance beam analogy of an open differential. Note that the total load at the fulcrum is equally divided between the pans.

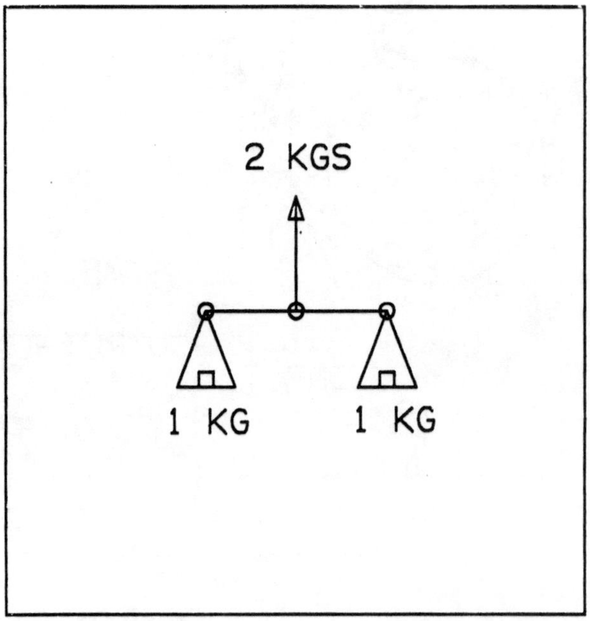

Figure 65
Balance Beam Analogy of Open Differential

Figure 66 represents an open differential. Like the balance beam in *Figure 65*, the input torque is equally divided between the two side gears by the pinion mate.

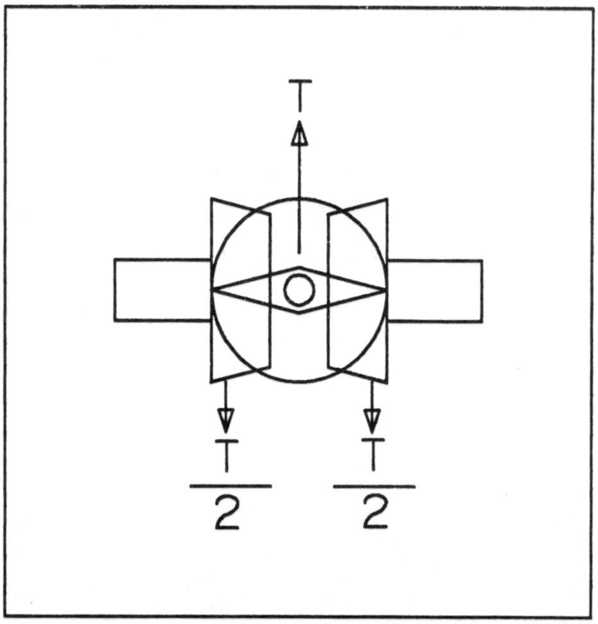

Figure 66
Open Differential

In *Figure 67* we see that the pinion mate becomes a continuous balance beam.

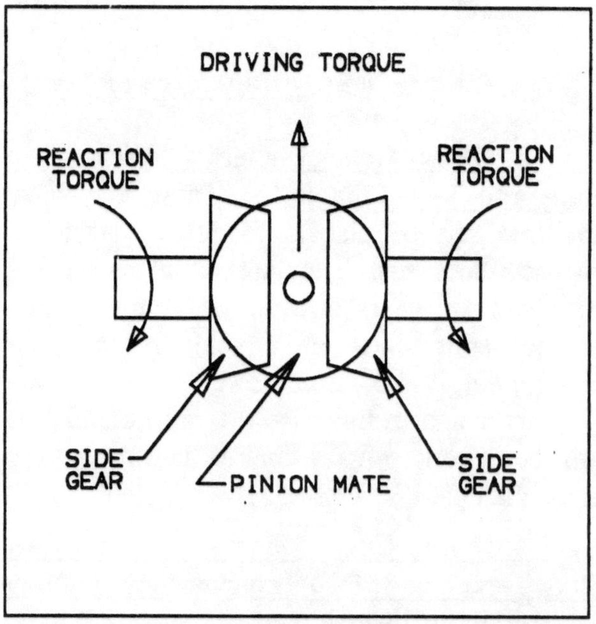

Figure 67
Open Differential

Locked Differential — *Figure 68* is a balance scale with the fulcrum locked to prevent rotation at the knife edge. With the pivot locked, weight can be placed on either pan and the total force will be supported at the fulcrum.

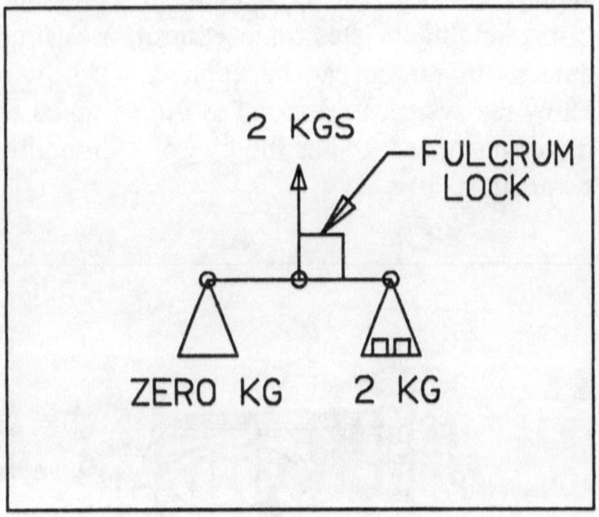

Figure 68
Scale with Fulcrum Locked

Figure 69 represents a locked differential. Note that the total input torque can be transmitted to either side gear and thus to either output. This is exactly like locking the pivot on the balance scale of *Figure 68.*

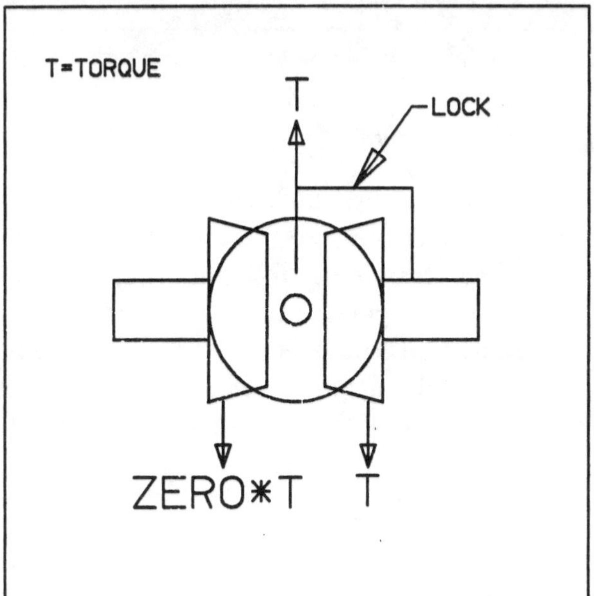

Figure 69
Locked Differential

DEVICES TO RESTRICT DIFFERENTIALS — Many different types of devices have been employed to partially restrict or totally restrict (or lock) differential action. Their function is to be able to apply more torque to one output. An example of the application would be when one wheel of an axle is on a slippery surface and the other wheel has good traction.

1. **Sliding Tooth Collar** — The sliding tooth collar is the simplest and probably the most economical way to restrict differential action. The collar is mounted on a splined shaft attached to one member of the differential and moved along the splined shaft with a shift fork. In one position it is disengaged and in another position the teeth interlock with an appropriate set of teeth on another member of the differential to prevent relative rotation. (See *Figure 70*)

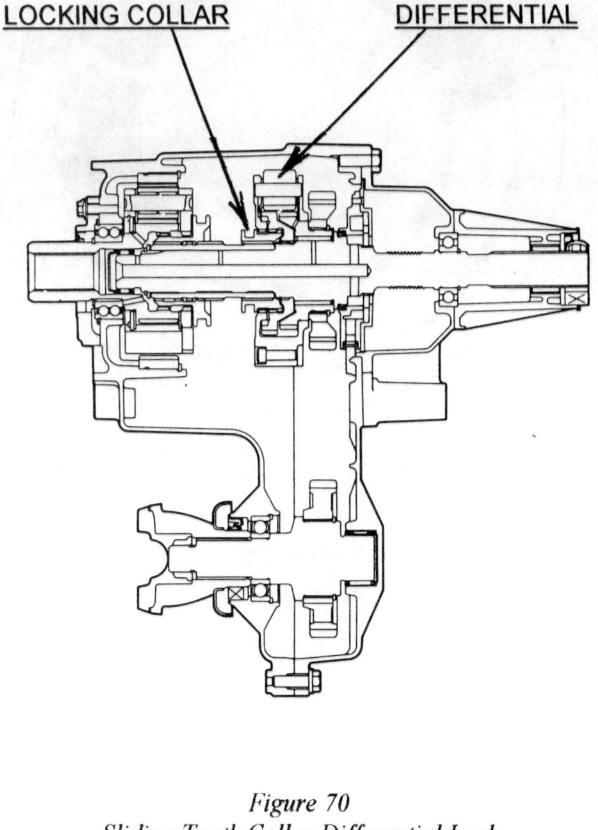

Figure 70
Sliding Tooth Collar Differential Lock

2. **Sprag Clutch** — A sprag clutch (*Figure 71*) operates much like a ratchet except it does not have finite steps. It has an inner and an outer race similar to a roller bearing. The sprags are positioned between the inner and outer race much like the rollers of a bearing. Unlike rollers, the sprags are elongated and set at an angle. They are contained by a retainer and are usually spring loaded to stay in contact with the inner and outer race. With the sprags set at an angle, relative rotation of the races in one direction will wedge them into place and prevent rotation. Turning the races in the opposite direction simply rotates the sprags to a free position and the races turn freely.

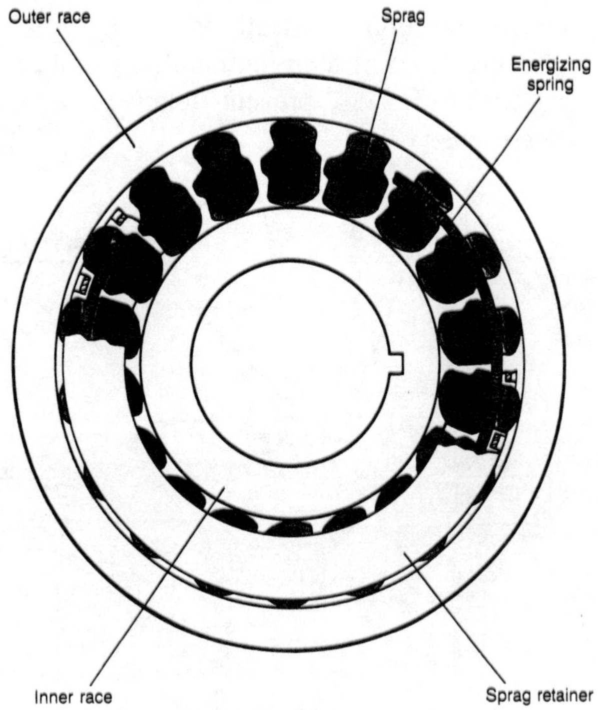

Figure 71
Sprag Clutch

3. **Hydraulically Actuated Multi-Disk Clutch** — A multi-disk clutch (*Figure 72*) activated by force from a hydraulic piston has also been used as a differential restricting device. This can be a very versatile and effective mechanism. An external hydraulic supply can be used as a source of pressure. Using intelligent electronic control, a desired amount of torque can be applied. This will allow the system to respond to programmed or driver-instigated handling or mobility commands.

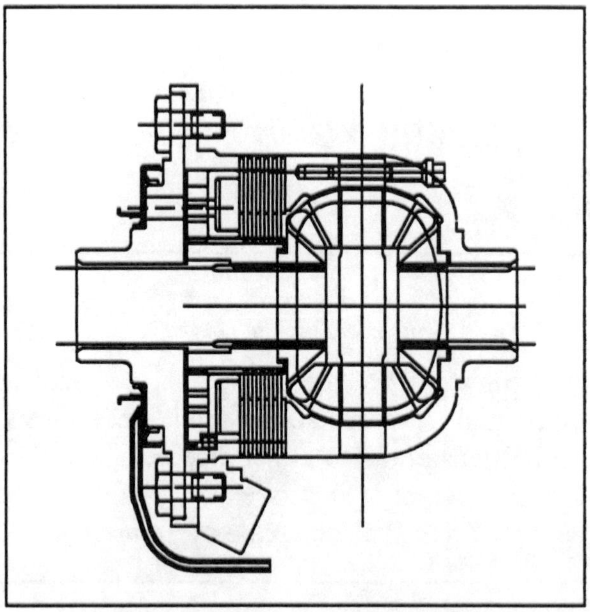

Figure 72
Hydraulic Multi-Disk Clutch with Internal Pump

This type of clutch can also be supplied with hydraulic pressure from an internal pump driven by the speed difference between the differential's input and one output, or between the differential's outputs. In this case, pressure and the resulting torque are a function of the amount of differential speed, i.e. the slip velocity.

4. **Torque Actuated Multi-Disk Clutch** — A bevel gear differential is an ideal place to put a multi-disk clutch for activation by the transmitted torque. Gear separating forces resulting from transmitted torque push the side gears outward into the differential case. Placing a clutch pack between the side gears and the case results in using the gear separating force to activate the clutch, thus restricting differentiation. The clutch restricting force or torque is a linear function and in direct proportion to the torque being transmitted by the differential. *Figure 73* is representative of this type of device.

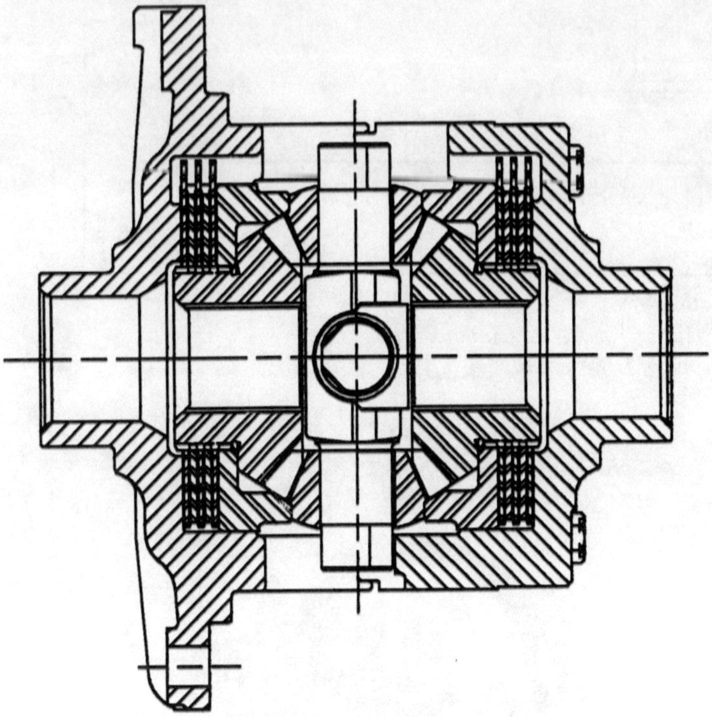

Figure 73
Bevel Gear Differential with Clutch Disks Placed Between Side Gears and the Case.

Placing a belleville spring within the clutch pack produces a preloaded clutch. (*Figure 74*) With preload springs in this position the differential will be restricted from rotation even with no transmitted torque. As torque is applied, resistance to differentiation will remain constant up to the point where spring preload is equal to gear separating force. From that point on clutch torque will increase in direct proportion to transmitted torque.

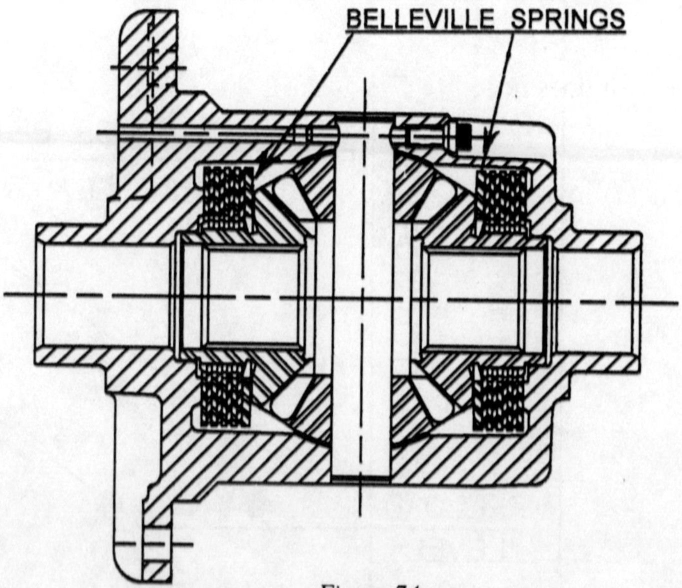

Figure 74
Differential with Belleville within Clutch

Placing springs between the side gears also will produce a preloaded clutch. In this case, however, the preload is in addition to the separating force. Clutch torque will be at a given level at zero torque and rise linearly as transmitted torque increases. *Figure 75* is an example of this type of differential.

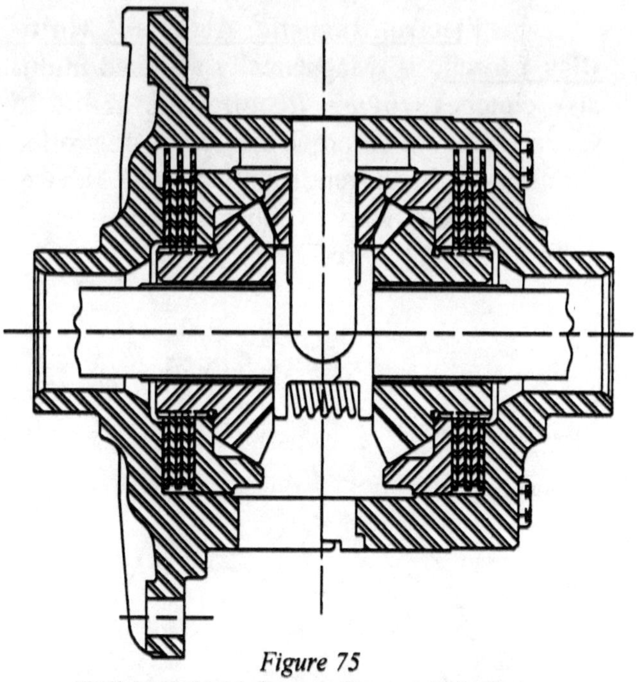

Figure 75
Differential with Springs Between Side Gears

Figure 76 depicts the clutch torque curves for the three differentials discussed above.

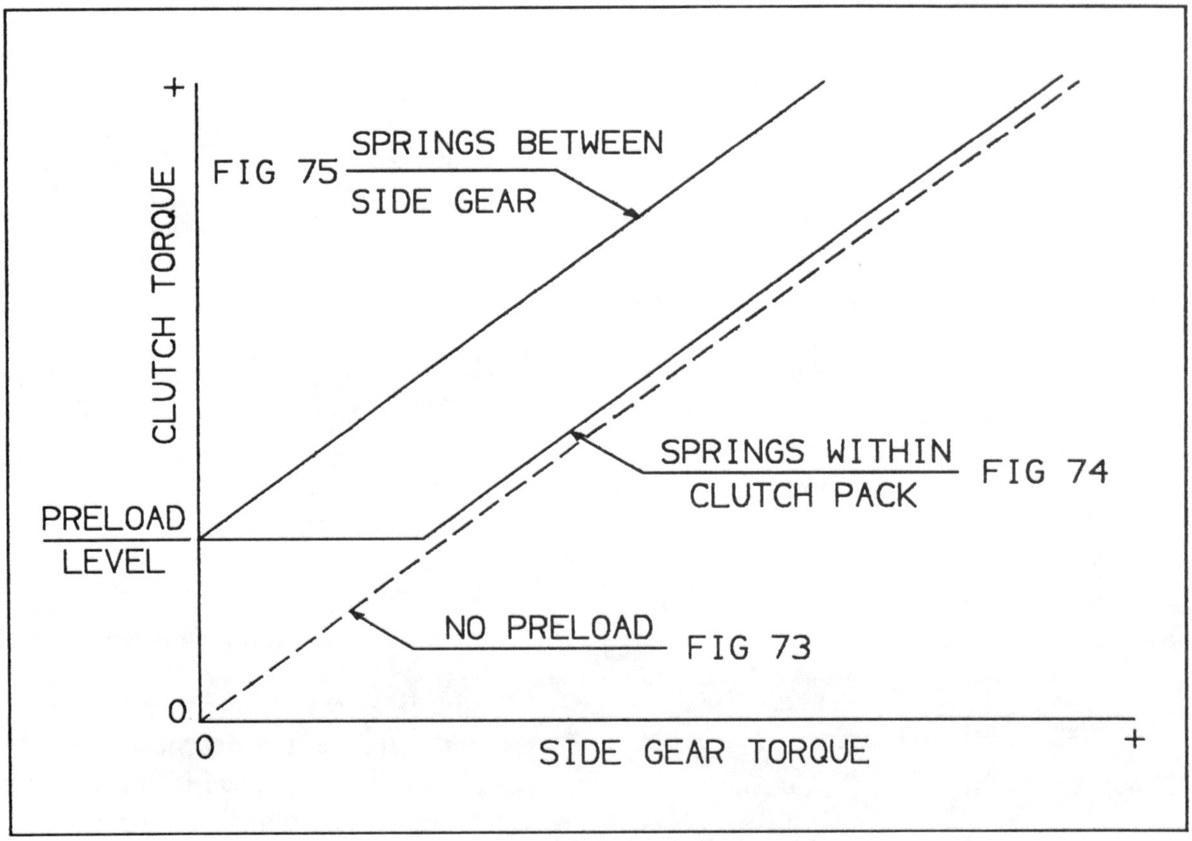

Figure 76
Composite Curve Showing the Transmitted Torque vs. Clutch Torque Curves of the Three Types of Differentials

5. **Electro-Magnetic Actuated Multi-Disk Clutch** — Magnetically actuated multi-disk clutches (*Figure 78*) offer the ability to vary the amount of torque applied by controlled variation in the current supplied to the device. They respond quickly to either computerized or driver input. (*Figure 77*)

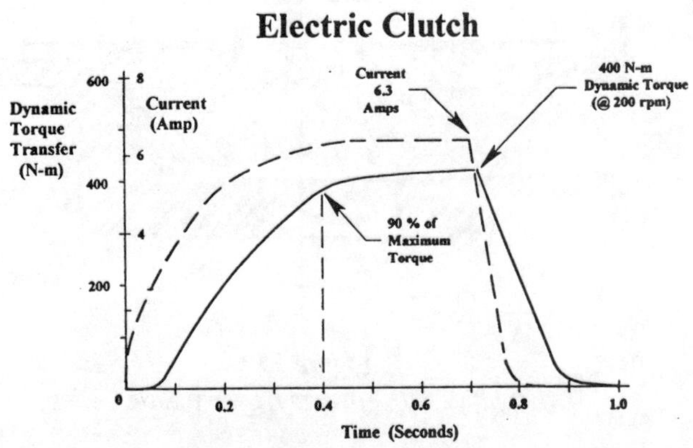

Figure 77
Electric Clutch Reaction Curve

Figure 78
Electro-Magnetic Inter-Axle Differential Restricting Clutch.

6. **Viscous Coupling** — The viscous coupling (*Figure 79*) is a multi-plate clutching device with close clearance between the plates and disks. It is usually sealed and filled to a specific level with a viscous fluid that presents high shear forces. Transmitted torque is a result of shearing the fluid.

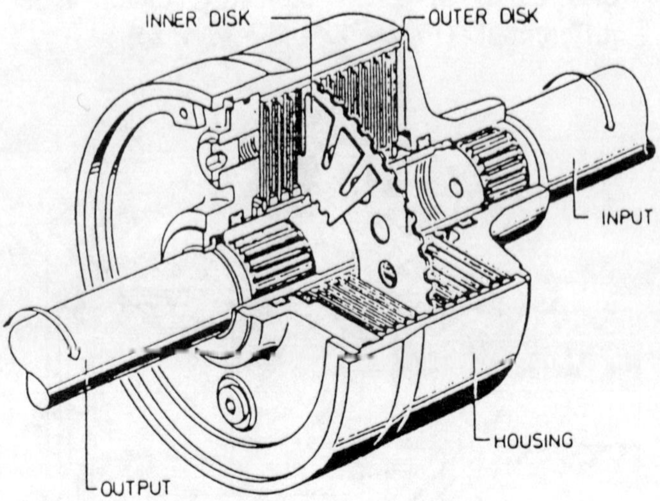

Figure 79
Viscous Coupling

Torque level increases as angular velocity between output and input increases. The torque curve tends to rise to a maximum level asymptotically. As heat is generated within the device, the fluid expands and the shear forces go up as a result of the pressure and temperature. This produces a condition known as "humping."

The term "humping" is derived from an anomaly that occurs when plotting torque curves. As temperature goes up, the amount of torque transmitted at a given angular velocity is greatly increased. This produces a "hump" in the speed vs. torque curve — thus the term "humping." Examples of viscous coupling curves are shown. (*Figures 80-a and 80-b*)

The shaded band in *Figure 80-a* is typical of variations that may occur as a result of mechanical tolerances and fluid viscosity variation.

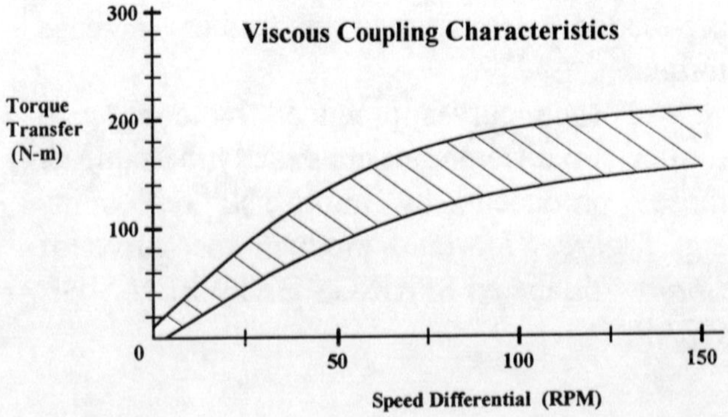

Figure 80-a
Torque Range vs Speed Curve

Typically a viscous coupling will require 15 to 25 seconds at 100 rpm to produce enough heat to go into the hump mode. (*Figure 80-b)*

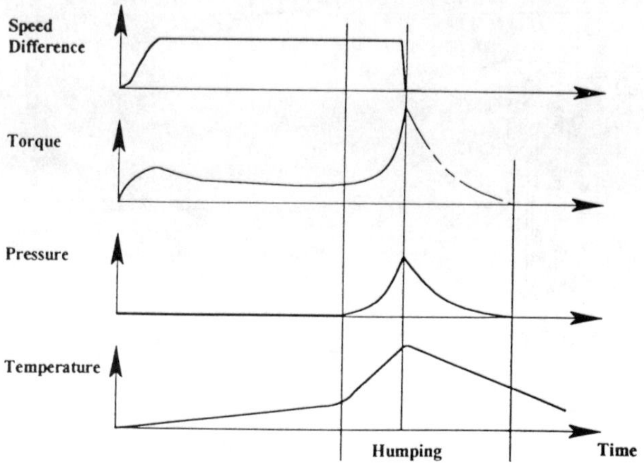

Figure 80-b:
Curves Showing Speed, Torque, Pressure and Temperature to Produce Humping.

7. **Cone Type Clutch** — Cone clutches are applied to bevel gear differentials. The cones are placed between the side gears and the case. Separating forces from the side gears under torque force the cones into the seat creating clutch torque. The cones can also be pre-loaded with springs to further increase torque..

Torque curves produced by cone type limited slip differentials are exactly the same as those produced by multi-disk type units, e.g. *Figure 81* would produce the curve of *Figure 76* labeled SPRINGS BETWEEN SIDE GEAR.

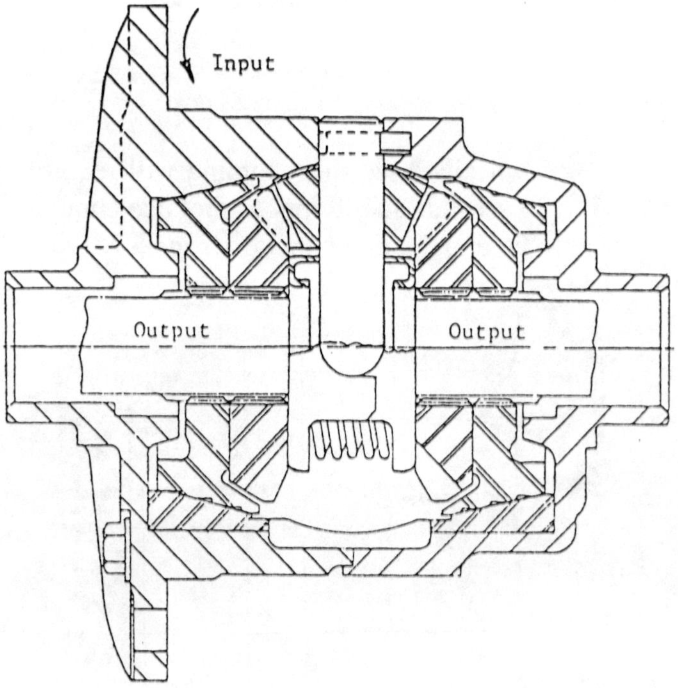

Figure 81
Loading Cone Clutch in Differential

Some cone clutches have been designed so they are preloaded and side gear torque tends to release the clutch. These are known as unloading cone types.

8. **Unconventional Differential Devices** — Worm gear type differentials (*Figure 82*) and various sliding element differentials are designed to take advantage of the inherent inefficiency of the device in order to prevent free differential action. Gear helix angles and sliding cam block angles are employed and designed in to give the desired amount of friction. These devices have the same torque bias curve as a non-preloaded clutch type differential. (*Figure 73* and *Figure 76*)

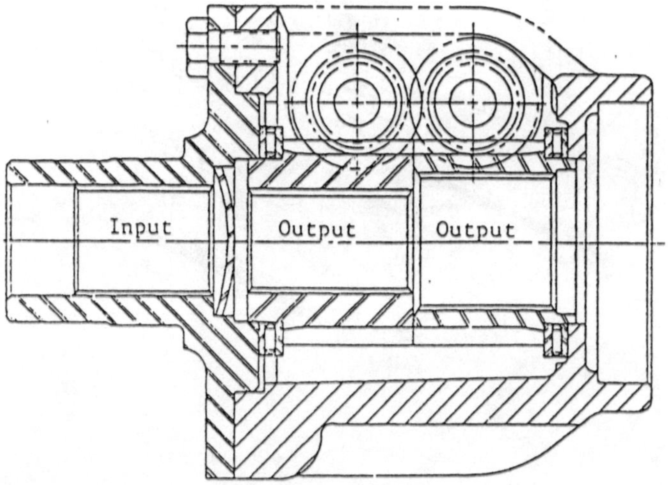

Figure 82
Worm Gear Type Differential

TORQUE DISTRIBUTION — The effect on torque distribution of several locking devices will be discussed:

A. <u>Positive Locking Devices</u> — These devices totally restrict differentiation, thus causing both outputs of the differential to rotate at the same velocity.

1. **Sliding Tooth Collar**

A tooth collar that joins any two rotating members of a differential device is a positive lock that prevents any differential action. Because of this, torque output can be up to 100% of input at either output leg.

2. **Sprag Clutch**

When engaged, a sprag clutch forces both members to turn as a unit just the same as a tooth collar. The sprag may allow lock in one direction so that one shaft can overrun the other. In this case torque will only be transmitted by the sprag when there is a specific direction to the relative motion between the shafts.

3. **Multi-Disk Clutch**

(a) Hydraulically Actuated

(b) Electro-Magnetically Actuated

Hydraulically actuated and electro-magnetically activated clutches can be used as positive locking devices if the clutch capacity is designed to be above wheel slip torque.

B. <u>Limited Slip Devices</u> — Limited slip devices are mechanisms inserted in the drivetrain in such a way as to partially restrict differential action. They are a compromise between an open and a fully locked system. Allowing differentiation, but requiring a degree of torque to achieve it, can increase tractive effort in some situations.

As described earlier in the paper, restricting differential action can produce a problem when the vehicle is turning a corner. If the differential is completely locked, the vehicle may hop or skip around the corner as torque is alternately wound up in the drivetrain and released as the tires hop or skip on the pavement. If the differential is partly locked by a multi-disk clutch, the clutch may intermittently release as torque builds up, producing a chatter or jerking effect in the drivetrain and making noises that are unacceptable to the driver. For this reason these devices must rotate freely, restricting torque without chatter. Special lubricant additives are often used to prevent the unwanted chatter. Many different clutch plate friction materials are available and some are chatter resistant, depending on the sensitivity of the system.

1. **Multi-Disk Clutch**

(a) Hydraulically Actuated

(b) Electro-Magnetically Actuated

(c) Torque Actuated

Multi-disk clutches have a finite amount of torque that can be transmitted before they begin to slip. This torque level can be adjusted by varying plate pressure, or normal force, and by the number and size of the disks employed. If they are designed so the plate pressure can be varied, the amount of resistance to differentiation can be controlled. If continuous slipping of the clutch is desired for a particular control situation, the clutch will need to be designed to withstand the heat generation that will result. Larger plates, circulated cooling oil etc. may need to be employed.

2. **Cone Type Clutch**

Cone clutch function is nearly identical to multi-disk clutches so the same techniques can be applied.

3. **Viscous Couplings**

Viscous devices also have a defined amount of torque dependent on size, number of plates, type of fluid, fill level, and the relative angular velocity between the plates. This torque variation can be used to achieve various desired performance results. Setting the viscous torque level too high can result in interference with an anti-lock braking system.

Since viscous couplings must slip to achieve torque there is a tendency for them to build heat. As they get hot they tend to go into a humping mode and the torque level goes up sharply. This characteristic will sometimes save them from complete destruction. A knowledge of the mechanics of humping can be very valuable to the system designer.

4. **Unconventional Differential Devices**

Worm gear devices have the same characteristics as bevel gear differentials with multi-disk non-preloaded clutch packs or non-preloaded cone type clutches.

PLACEMENT OF CLUTCHING DEVICES — With variable slip devices it is important to recognize the effect on output torque when the clutching device is placed to operate between different differential members. Placement options using one clutch are:

- Between the input and one output.
- Between the input and the other output.
- Between the two outputs.

More than one clutch can be employed in the same differential, usually with each operating between one output and the input. This is typical in rear axle limited slip differentials.

Bevel Gear Differentials — The clutch placement options for a bevel gear differential are:

1. **Clutch placed between input and one output.** (*Figure 83*) In this arrangement the amount of torque delivered to one output can be different than that delivered to the other. Due to gear separating forces, a left wheel slip can generate more clutch torque than a right wheel slip.

With zero torque at the left axle torque to the right axle passes through the gears from the clutch and gear separating forces come into play. Zero torque at the right axle allows torque to flow directly through the clutch to the left axle and no gear separating force is generated.

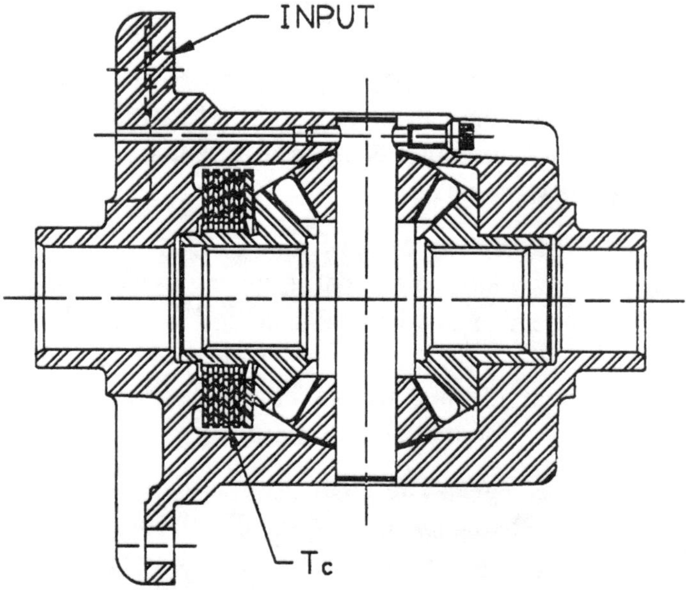

Figure 83
Clutch Between Input and One Output

2. **Two clutches, each placed between the input and one output.** (*Figure 84*) In this arrangement the output torques are balanced.

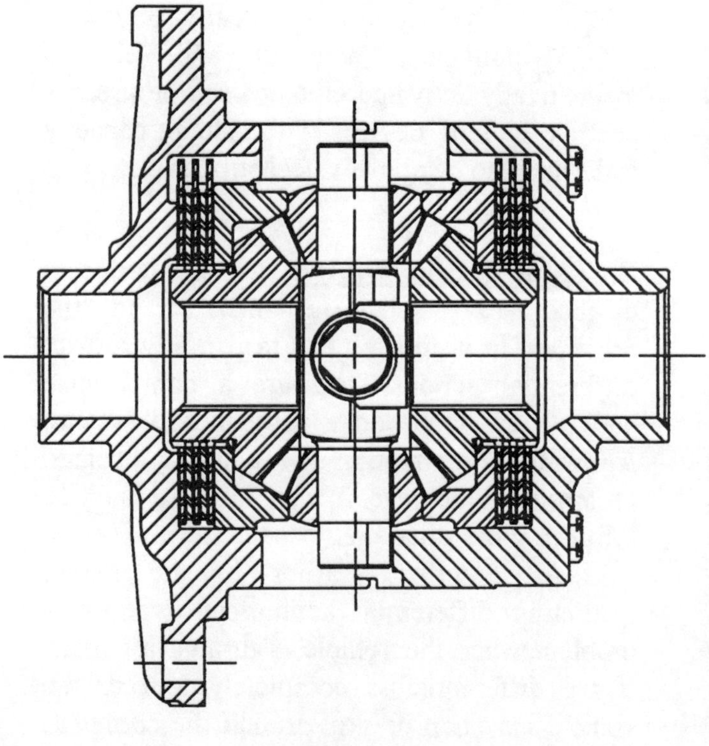

Figure 84
Two Clutches, One at Each Output

Planetary Gear Differentials — Carrier input planetary differentials used in transfer cases offer the most interesting clutch placement options. This is due to the rather large torque split capabilities possible. Analysis of this type should serve the designer as a warning to carefully analyze the effect of recirculating torque in any differential device coupled with a clutching mechanism to partially restrict differentiation. Three clutch placement options are analyzed:

1. **Clutch placed between ring gear and sun gear (between outputs)** (*Figure 85*)
2. **Clutch between the sun gear and carrier (input and one output)** (*Figure 86*)
3. **Clutch placed between ring gear and carrier (input and one output)** (*Figure 87*)

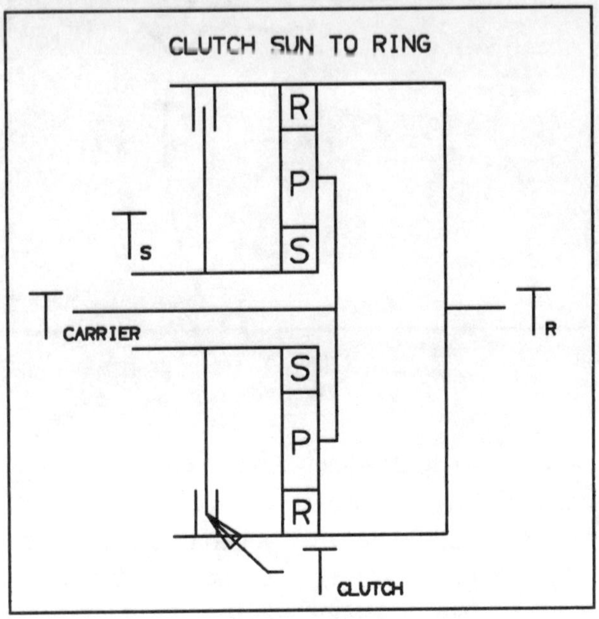

Figure 85
Clutch Between Outputs

When Ring Gear is at 0 Torque:

$$STQ = CLTQ + CLTQ * \frac{SG\ Teeth}{RG\ Teeth}$$

When Sun Gear is at 0 Torque:

$$RTQ = CLTQ + CLTQ * \frac{RG\ Teeth}{SG\ Teeth}$$

KEY TO EQUATION SYMBOLS

RTQ = Ring Torque   SG Teeth = Sun Gear Teeth
STQ = Sun Torque   RG Teeth = Ring Gear Teeth
CLTQ = Clutch Torque

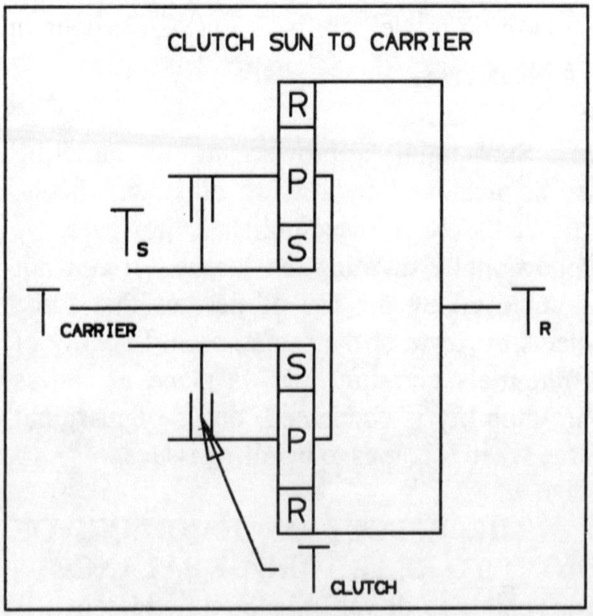

Figure 86
Clutch Between Sun and Carrier

When Ring Gear is at 0 Torque:
Sun Torque = Clutch torque = Carrier Torque

When Sun Gear is at 0 Torque:

$$\text{Ring Torque} = \frac{\text{Clutch Torque} * \text{Ring Teeth}}{\text{Sun Teeth}}$$

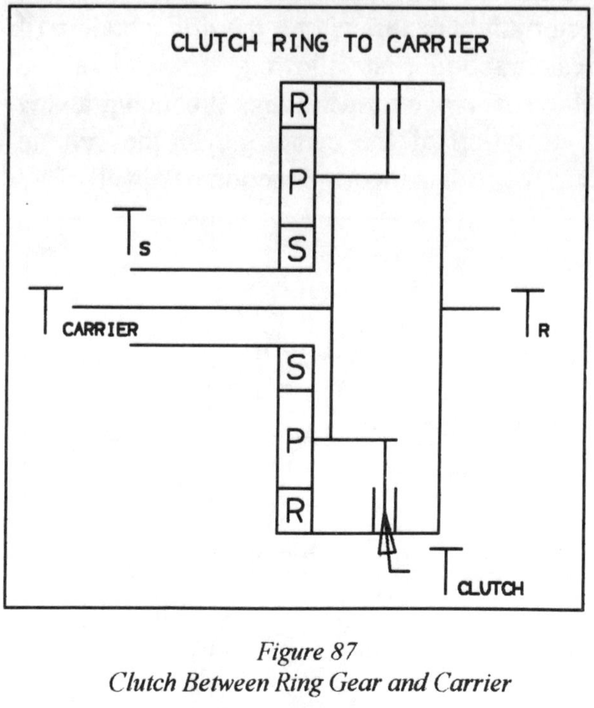

Figure 87
Clutch Between Ring Gear and Carrier

When Sun Gear is at 0 Torque:
Ring Torque = Clutch Torque = Carrier Torque

When Ring Gear is at 0 Torque:

$$\text{Sun Torque} = \frac{\text{Clutch Torque} * \text{Sun Teeth}}{\text{Ring Teeth}}$$

HANDLING

Substantial improvements in handling can be achieved by driving all four wheels. This can only be accomplished, however, by proportionally dividing the torque. It can not be achieved by the use of devices that force wheels to rotate at the same speed. Locking of differentials certainly has its place as far as pure mobility is concerned, but is of marginal value when it comes to handling issues.

THE EFFECT ON HANDLING OF FRONT TO REAR TORQUE BALANCE — If a vehicle is driven in a circle and speed is increased, the amount of centripetal force required to keep the vehicle on a circular path will increase. Eventually this force will approach the tractive capability of the tires.

On the driving axle of a two-wheel-drive vehicle the forward propelling force and the lateral or side skid force are additive vectorially. For this reason the driven axle will break traction first, allowing that end of the vehicle to move outward thus producing a yaw or deviation of the centerline of the vehicle from the instantaneous direction of travel.

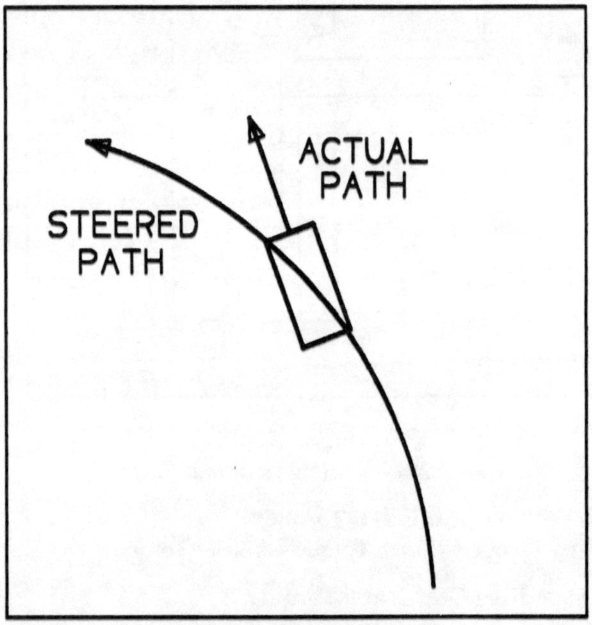

Figure 88
Understeer

In a front drive vehicle the front tends to skid out causing the turning radius to become larger than called for by the steer angle. This condition is known as *understeer*. (*Figure 88*)

In a rear drive vehicle the rear skids out first and now the vehicle points inward trying to produce a smaller turning radius than called for by the steer angle. This condition is known as *oversteer*. (*Figure 89*)

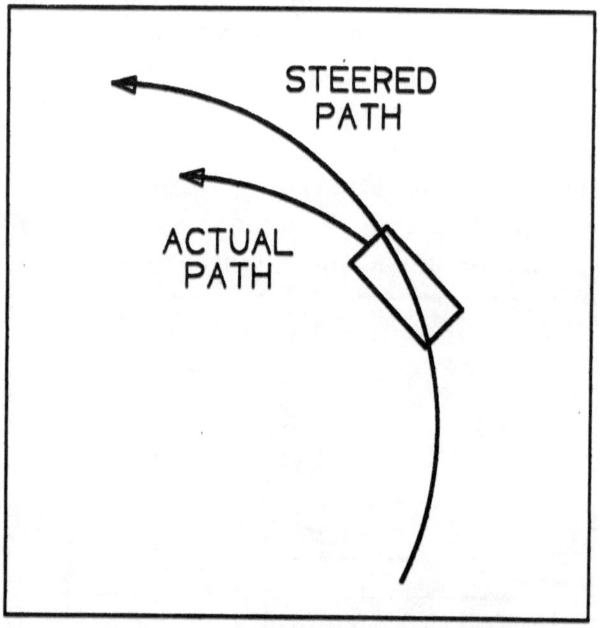

Figure 89
Oversteer

In a four-wheel-drive vehicle with a center differential we have the ability to affect the oversteer/understeer characteristics. The vehicle can get around a corner faster without yawing in an unwanted manner. (Vehicle tests should be run to be sure the vehicle won't break away in an uncontrollable manner.)

Now that we understand that cornering performance can be improved, it becomes obvious that all-wheel-drive can be set up to give an all around better handling vehicle.

What about improving handling by the addition of a differential restrictor placed across the inter-axle differential? A relatively instant acting system like an electronically controlled magnetic clutch or an electronically controlled hydraulic clutch will still take about 300 milliseconds to achieve a significant amount of lock. This will allow the slipped wheel to rotate relative to the point of contact by 1/2 to 1 full turn and will require about 13 feet of travel at only 30 mph. Since the skidding out process is well under way any improvement in handling is dubious.

A locking differential in the rear axle of a full-time 4-wheel-drive vehicle will produce the same result as a locking differential in a 2-wheel-drive rear drive vehicle. Both rear wheels will tend to break away at the same time. (Race drivers like this: they can set up a controlled rear skid by the amount of throttle applied, and then they can turn into the skid to adjust the cornering path of the vehicle.)

From the above we can conclude that the greatest effect on handling is achieved with an open differential. Only marginal improvement can be achieved by adding clutching devices.

Viscous type devices require a high relative angular velocity to produce useful torque levels. The wheel slip must continue if any torque is to be transmitted to a non-slipping wheel. This is unproductive, since the desired object is to arrest wheel spin-up.

It is obvious that the act of locking the differential will have little if any advantage in handling and may even be detrimental.

An ideal system for improving handling would be one that contained a variable torque inter-axle differential. Then if a wheel begins to spin-up, the torque balance of the differential could be changed to provide more torque to the non-spinning wheels. This is not the same thing as locking or restricting differential action. In our balance beam analogy, this would be the equivalent of shifting the fulcrum on the balance beam toward the non-slipping wheel. A locking device would be the equivalent of locking the fulcrum of the balance beam.

FUNDAMENTALS OF HANDLING —

Tracking — Tracking is the ability of the vehicle to maintain the direction of travel desired by the driver. A driven wheel is less prone to being deflected off course by variations in the road surface. An example of this is the extreme condition of driving in deep ruts. A non-driven steering wheel will skid along the rut even when substantial steering input is applied. A driven steering wheel will have a tendency to climb out of the rut, thus responding to the drivers input. This condition remains in a more subtle manner on improved roads and the driver is left with a feeling of far greater directional control than in an ordinary vehicle.

Stability — Variations in the road surface will cause greater deflection of the vehicle from the intended path in an undriven steering wheel. Take the case of crossing a railroad track at an angle. The horizontal component of the impact will cause a steering deflection, requiring driver input to correct. The result is a swerve and loss of a stable feel in the vehicle. If the wheel is driven it will have a tendency to climb the rail and the effect of the horizontal component will be reduced, resulting in less need for driver input, less oscillation of the vehicle and a better and more stable feel.

Cornering — The effect on cornering was illustrated in an earlier section on Oversteer/Understeer. Proper balance of torque front to rear results in minimizing both oversteer and understeer characteristics. This allows the vehicle to negotiate the corner at a higher speed or simply better maintain the desired radius at lower speeds. This is due to a reduction of the tendency for one end of the vehicle to slip outward as a result of using up the total available traction at that set of wheels.

Acceleration and Performance — Consider a two-wheel-drive vehicle with a small portion of its weight on the driving axle. With good engine torque and a high numeric drivetrain ratio, the driving wheels will not be able to absorb all the engine's power. Acceleration capability will be lost.

A full-time system will divide the power among all four wheels tending to prevent wheel slip. A far greater rate of acceleration will be achieved, and the desired speed reached quickly without wheel spin. It could be argued this will enhance stability as well as increasing performance.

In pick-up trucks, powering all four wheels will give better towing capability. It will also provide a lower continuous torque level to the rear axle. This could allow a one model smaller ring gear, pinion and differential thus increasing vehicle efficiency and load capacity.

Hydroplaning — When it is raining and there is a film of water on the pavement, tires can ride up on that film and begin to skid. Since the phenomenon is much the same as a water skier or boat skimming over the water it is known as "hydroplaning".

When a wheel rides up on the water film and begins to skid, if it is a non-driven wheel it may even come to a stop. A driven wheel will tend to keep the contact area at 0 velocity relative to the road, and thus will cut through the water film just as if the vehicle were standing still. Better directional stability on wet surfaces is one of the many advantages of driving all four wheels.

Traction — Four wheels with even loading on a given surface can provide double the traction of two. Many vehicles have the power to easily break traction of two wheels. Spreading that power to more wheels, will prevent break-away and tire squealing. Rapid acceleration will be achieved without loss of control.

DRIVETRAIN SYSTEMS — A lot has been written about all-wheel-drive systems and four-wheel-drive systems and their ability to enhance vehicle handling. In analyzing these systems we find some very interesting physical situations. These arise as a result of the differences in the drive systems employed.

The different types of four-wheel and all-wheel-drive systems are:

1. **All-Wheel-Drive Systems** —

- System has a center differential.
- The center differential divides torque front to rear: 50/50, 60/40, 30/70, etc.
- Center differential is restricted from free rotation and the resulting torque division by a clutching device operating between two of its members. The clutching device may partially or wholly prevent differentiation.

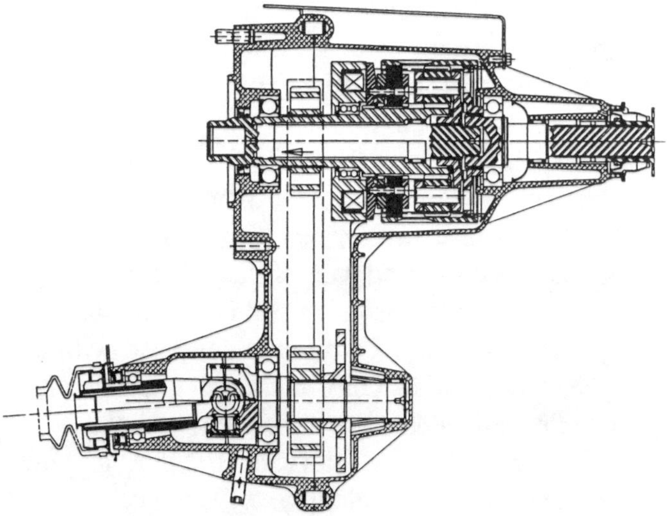

Figure 90
Cross Section of an All-Wheel-Drive

2. **Part-Time 4-Wheel-Drive Systems** —

- Usually 2-wheel rear drive with a clutching device to drive the other axle.
- Clutch is usually positive, causing front and rear propshafts to turn at the same speed.

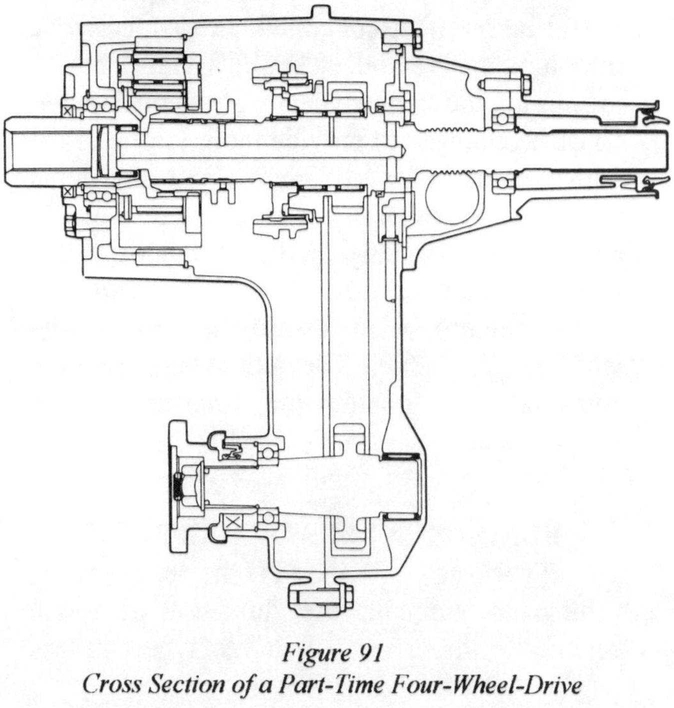

Figure 91
Cross Section of a Part-Time Four-Wheel-Drive

3. **<u>On-Demand Systems</u>** —

- Cannot divide torque.
- Have no center differential.
- Systems have a variable clutching device to transmit power to the normally inactive axle when wheel slip is detected at the main driving axle.

4. **<u>Continuously Variable Systems</u>** —

- Fixed torque ratio of the center differential is replaced by a variable torque ratioing system.
- Torque ratioing is done by a continuously variable device such as a variable V-belt, toroidal drive or variable hydraulic drive.

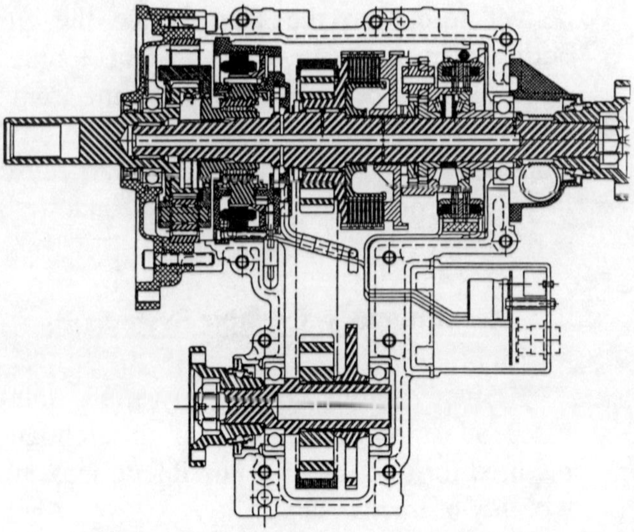

Figure 92
Cross Section of an On-Demand Four-Wheel-Drive

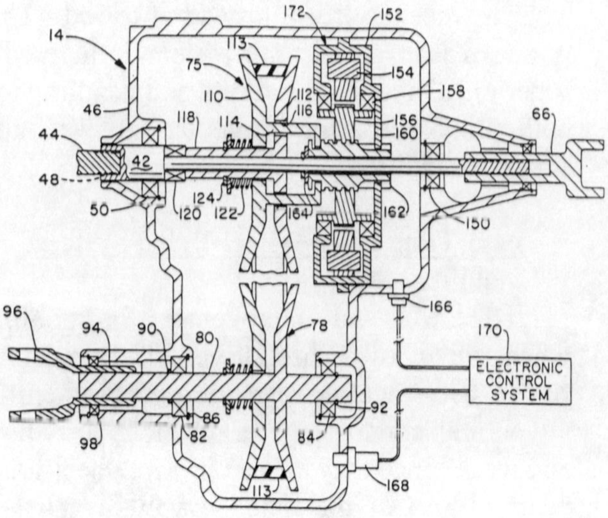

Figure 93
Cross Section of a Continuously Variable Four-Wheel-Drive

ANALYSIS OF SYSTEMS — Listed below are comparisons of mobility and handling of the various all-wheel and 4-wheel drive systems.

<u>The All-Wheel-Drive System</u> —

- Mobility:

Mobility is improved over a two wheel drive system in direct proportion to the differential torque split ratio and the effect of weight distribution of the vehicle, e.g. a 50/50 torque split with 50/50 weight distribution would give maximum tractability with like coefficients of friction under each wheel. Mobility or tractability can be further enhanced by having a differential restricting device. Maximum tractability is, of course, achieved by positive lock of the differential. This is useful when different coefficients are encountered or weight distribution is unequal.

- Handling:

The continuous delivery of torque to all four wheels can significantly improve stability and cornering capability of the vehicle. Proper balance of torque front to rear can virtually eliminate the tendency to oversteer or understeer. Further attempts to enhance cornering by partially locking the differential may or may not be beneficial. As differentiation is restricted, a locked system is approached. Since the front overruns the rear, locking the differential brakes the front and adds torque to the rear. A viscous device tends in this direction until a rear wheel slips out, then as the wheel spins faster torque is increased to the front. An electronically controlled device will not interfere with torque transfer until a wheel slips and will then transfer torque to the degree of lock called for by the program. It can be controllable and not dependent on wheel slip velocity.

Hydroplaning is significantly reduced when a wheel is driven instead of towed. This is one reason a full time system is markedly superior. Directional stability and handling are both improved when running in wet and slippery conditions.

Part-Time Four-Wheel-Drive-Systems —

- Mobility:

There is no improvement in mobility when run in two-wheel-drive. They must be run in two-wheel-drive mode on pavement in all normal conditions. (Exceptions are snow and ice.) Since they are run in two-wheel-drive nearly 100% of the time, very little value is realized. Once in a while, when there is snow or ice on the pavement, people get to shift into four-wheel-drive and realize some value from all that hardware they have paid for and have been hauling around. Maximum tractability is available for off-roading and in the rare events when a locked four-wheel-drive system can be employed.

- Handling:

No improvement in handling is realized while running on-road, since 2-wheel-drive must be selected. In fact, if 4-wheel-drive is selected, handling is impaired by the drivetrain wind-up and the tendency of the vehicle to hop or skid around corners.

On-Demand Systems —

- Mobility:

Mobility is not as good as a full-time system since no torque is delivered to the front until slip is initiated at the rear. The system can be locked, giving identical off-road capability as a part-time system.

Pure tractability is nearly as good as either the AWD or the part-time system in the locked mode.

- Handling:

Handling is virtually the same as a two-wheel-drive system since no torque can be delivered to the front until slip is initiated at the rear. Partially engaging the clutch does not deliver torque to the front, since the front normally overruns the rear. In fact, torque is more likely to be transferred from the front to the rear, thus producing a braking action on the front and increasing torque to the rear, causing it to slip out even sooner (an undesirable condition).

Continuously Variable Systems —

- Mobility:

Since torque can be varied totally independent of the amount of differentiation required for steering, it is possible to maximize mobility or tractive capability.

- Handling:

With a continuously variable system, torque to the front and to the rear can be monitored and regulated as required to provide the ultimate in handling enhancement.

NOTE: All other systems either partially or totally lock the front and rear drives together, urging them to rotate at the same speed. In doing this they control only the amount of torque that restricts relative movement between the front and rear drives. Torque transmitted to either axle is dependent on steer angle and coefficient of friction at the tires, and may have an undesirable effect on handling.

TRACTION

Traction from the standpoint of pure drawbar capability will be influenced by the type of system and the amount of torque that gets to each wheel as a result of the system and coefficient of friction under each wheel.

TRACTIVE CAPABILITY — *Figure 94.* gives the tractive capability of a hypothetical vehicle with one wheel on a low coefficient surface and the other wheels on higher coefficients. Examining the table we find that four-wheel-drive always outperforms two-wheel-drive. **Not surprising!** (*Figures 95, 96, 97, & 98*)

TRACTION CURVES

COEFFICIENT		ALL–WHEEL–DRIVE		TWO–WHEEL–DRIVE	
Low μ Wheel	High μ Wheel	Without Differential Lock kilograms	With Differential Lock kilograms	Without Differential Lock kilograms	With Differential Lock kilograms
0.1	0.1	200	200	100	100
	0.2	200	300	100	150
	0.4	200	500	100	250
	0.6	200	700	100	350
	0.8	200	900	100	450
0.2	0.2	800	400	200	200
	0.4	800	700	200	300
	0.6	800	1000	200	400
	0.8	800	1300	200	500
0.4	0.4	800	800	400	400
	0.6	800	1100	400	500
	0.8	800	1400	400	600
0.6	0.6	1200	1200	600	600
	0.8	1200	1500	600	700

Figure 94

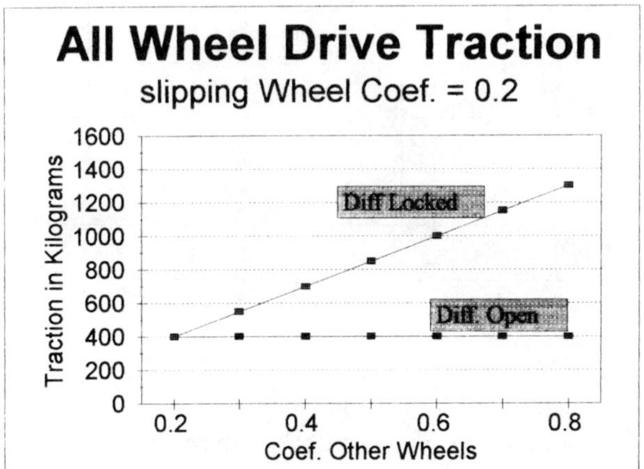

Figure 95

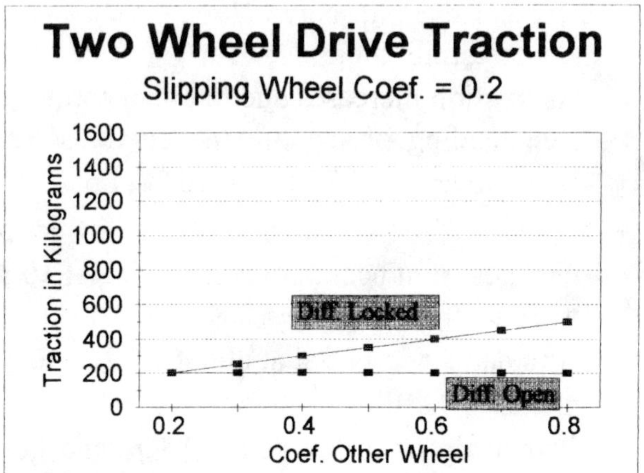

Figure 96

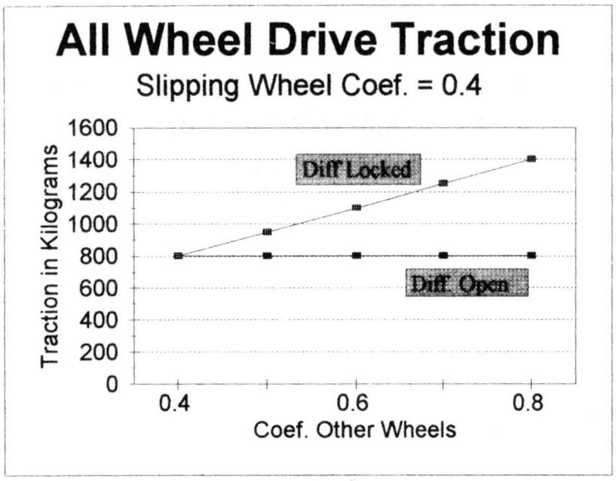

Figure 97

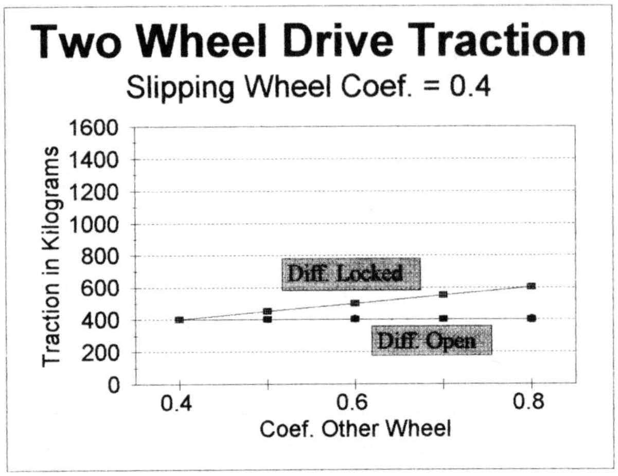

Figure-98

A common misconception about a three-differential four-wheel-drive system is that, if any one wheel spins out, it will absorb all the power and you will be worse off than with a two-wheel-drive system. **This is absolutely wrong!** As discussed earlier in the balance system analogy, the other wheels will receive torque in direct ratio to the differentials. If all three differentials are 50/50 units then we will have four times the slipping wheel torque to produce drawbar.

Further examination of the data and graphs shows how to predict exact vehicle tractive capability in a variety of situations.

TIRE μ VS. SLIP CURVES — A look at a typical tire slip curve (*Figure 99*) reveals some interesting things:

- As traction increases due to either torque or side loading of the tire the curve is near linear to about 0.75 μ and 10 % slip.
- From 0.75 μ to about 0.95 μ the slope changes from near vertical to near flat 10 % to 30 % slip.
- It remains nearly flat at about 1.0 μ out to about 90 % slip.
- It then drops off to about 0.8 μ and goes into 100 % slip.

Knowing where the vehicle is operating on the slip curve can tell a great deal about drawbar pull, oversteer, understeer, stability, cornering, etc. There have been computerized vehicle handling simulation programs written that can be used to predict a wide variety of handling situations. A knowledge of tire tractive capability can be useful data to insert into these programs when trying to maximize a systems capabilities.

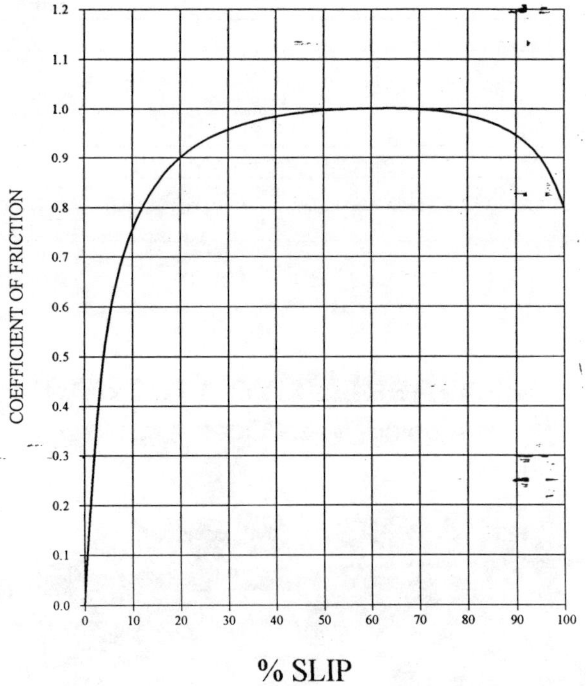

Figure 99
Tire Slip Curve, Dry Pavement

ABS TRACTION CONTROL — The use of Anti-skid Braking Systems (ABS) to enhance traction has become popular. If a wheel spins out, its brake can be applied to arrest the spin and additional torque can be applied to the wheel on the opposite end of the axle. Since the electronics and braking system are already in place this is a very economical way to enhance tractive capability.

Putting on the brakes to make a vehicle go not only seems inefficient, it really is. If the slipping wheel is brought back to road speed all the energy supplied to it, above that required to achieve skid torque, will be dissipated as heat. If the coefficient at the slipping wheel is near zero the available tractive force drops by nearly 50%. Additional engine power will be required to maintain the same tractive effort as before the slippery surface was encountered.

If the wheel is brought to a complete stop available tractive effort is less than 50% since the wheel now has to be pushed ahead by the opposite wheel.

DIFFLOCK TRACTION CONTROL — Applying a clutch to restrict differential action results in minimal energy loss. As soon as the differential becomes inactive the clutch ceases slipping and there is no more heat loss. Torque is fed back to the wheel with good traction and there is no need to increase engine power to maintain the same amount of tractive capability.

It's simple: don't put on the brakes to make a vehicle go.

OFF-ROADING — The off-roading capability of a three-differential all-wheel-drive system has been largely ignored. Off-roading has been considered the domain of fully locked four-wheel-drive systems.

Some interesting things have been discovered in recent off-road tests to compare various systems. One situation compared a locked (part-time) system vs. an open center differential (full-time) system doing tight turns and figure eights in a mud slurry about 20 cm deep. This test was done in a plowed field after several inches of rain. The full-time system continued to do both the tight circles and figure eights; the part-time system bogged down and got stuck. The front wheels of the part-time system are forced to run at the same speed as the rear. They need to go faster due to cornering. As a result they end up being pushed forward; and being set at an angle for steering, they plow into the mud, which piles up ahead of them, stopping the vehicle. This happened in both the circle and figure eight maneuvers.

The circles and figure eights were also tried on loose desert sand. The results were very similar to the mud scenario above. The full-time system would tend to climb up on top of the sand, much like a boat in plane on the water. The part-time system would tend to plow as it did in the mud. It did not get stuck but required a far greater amount of power and would not achieve nearly the speed of the full-time system.

Another desert run followed a dry creek bed with loose sand for a distance of several miles. When going straight ahead both systems would get up on plane and move right along. On corners the full-time system kept right on going; the part-time system tended to plow, requiring a lot more power and slowing down in the process.

Climbing the Rocky Mountain mining trails near and around Telluride, Colorado you see some of the most beautiful scenery in the world. Anywhere you point a camera you get a picture suitable for a postcard. You traverse some of the steepest, narrowest, rockiest, and most treacherous trails you might ever care to drive. Altitudes as high as 3962 meters are achieved.

Here the full-time systems would stay with the fully locked vehicles in 99% of all the situations encountered. The only situation that caused a need for locked operation was when the vehicle weight was supported by two opposite wheels as when climbing over large rocks or when going diagonally across a ditch. On a vehicle equipped with a rear axle diff-lock and an electronically controlled multi-disc clutch to lock the inter-axle differential, even these conditions were not a problem.

EFFICIENCY

Efficiency of four-wheel-drive vehicles is an often debated and usually misunderstood topic.

ENERGY LOSS IN COMPONENTS — The energy losses in drivetrain components are as follows:

- Stirring Loss – The splashing and churning energy required to move the lubricant around inside a gearbox as the gears and shafts rotate.
- Seal Drag – Energy lost to friction of the seal rubbing on a shaft.
- Bearing Drag – Normal friction to rotate a bearing.
- Gear Mesh Friction – The energy loss due to the finite sliding that occurs between gear teeth.
- Chain Friction – The normal losses that occur as a result of friction at the joints of a chain.
- Wind Losses – Energy loss due to stirring action of the air as when the yokes of a universal joint churns the air much as a fan. Also the wind loss that occurs due to various drive line components hanging in the slip stream under the vehicle.
- Rolling Resistance – Due to the added weight of the four-wheel-drive components there is a slight increase in the rolling resistance of the vehicle.

TIRE ENERGY LOSSES — *Figure 100* is a curve that shows the amount of energy loss in a tire at various load and driving conditions. It is interesting to note that the energy loss of a free rolling tire is somewhat higher than when the tire is driven. As torque is applied to a free rolling tire the energy loss decreases and then begins to increase as more torque is applied.

In a two-wheel-drive vehicle the driving wheels are usually to the right of the low point of the energy loss curve while the non-driven axle's wheels are at the free rolling point.

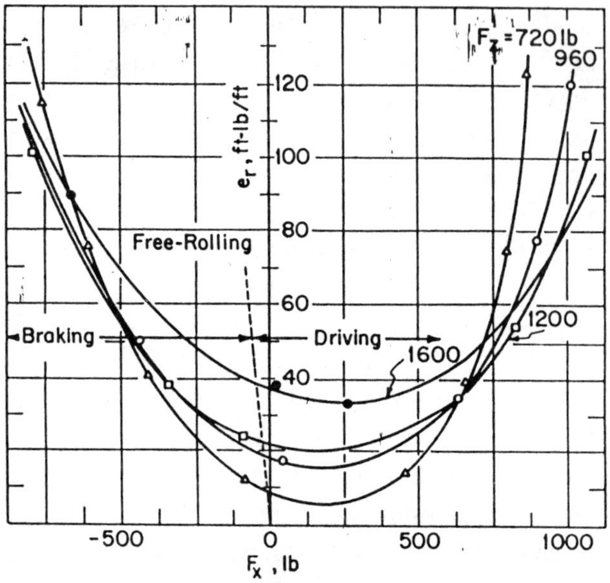

Figure 100
Effect of Longitudinal Force F_x and F_z on Energy Loss e_r For a 6.45S14 Tire at Zero Slip Angle

By driving all four wheels we can move the formerly non-driven axle to the right on its curve while the already driven axle moves to the left. There is a positive energy savings here that can be calculated and proven by testing.

ENERGY MANAGEMENT — Adding components to a vehicle to make it four-wheel-drive increases weight and wind resistance. The added weight and wind resistance tend to make the vehicle slightly less fuel efficient.

In an attempt to make four-wheel-drive vehicles more fuel efficient. many are set up to normally be run in two-wheel-drive. The idea is to eliminate the energy losses in the components driving the front axle. If the stirring losses, seal drag, and bearing friction are eliminated by disabling the front part of the drivetrain, there may be some energy savings. However, there is a better way to regain fuel efficiency.

Loss of energy due to stirring action is usually the largest contributor to drivetrain inefficiency. With the front wheels driven and stirring losses well managed, a vehicle can actually get better fuel mileage in full-time four-whcel-drive than in two-wheel-drive.

TESTING

LABORATORY TESTING — Testing four-wheel-drive components is no different than testing those for two-wheel-drive. Static, dynamic, and fatigue tests are required. Loads, speed and fatigue life, may be different depending on the intended use of a vehicle. With the large number of shafts, gears, seals, bearings, etc. a complete book could be written on component testing and the expected results. That will be left for a different time.

FIELD TESTING — Some of the special test requirements for all-wheel-drive systems will be discussed.

On-Road-Testing — On road testing is only slightly different than for standard two-wheel-drive systems. Exceptions are the need to test function of any traction enhancement systems especially those included to control an inter-axle differential. If a differential locking device comes on too quickly or with too much torque it may cause the vehicle to hop as it turns a tight corner. Friction clutches used in these devices may chatter, jerk. or produce undesirable sounds in cornering.

Handling and performance testing should also be run on high, low and split coefficient surfaces. Low μ will often indicate undesirable driver feel if too much clutch torque is applied across a differential. Split μ testing is useful to determine the effectiveness of traction enhancement systems.

Off-Road-Testing — Off-road testing of all-wheel-drive and four-wheel-drive systems is in general much more severe than for two-wheel-drive. Two-wheel-drive vehicles may not need to be off-road tested at all.

Some unique tests are:

- Sand Wash – Dry beach sand about 0.5 meters deep is spread over a hard clay base. The vehicles are driven in figure-eights and circles. This requires high torque to perform and is run at frequent intervals during durability tests.
- Log Climb Test – Logs or railroad ties are laid out in a random pattern so the left side of the vehicle and the right are climbing over different logs in an alternate fashion. This tests both the vehicle's structure and the ability to creep over obstacles.
- Steep Hills – Sixty to seventy percent slopes are negotiated both in climb and descent mode. Climbing ability and engine braking are important to serious off-roaders.
- Creek Beds – The ability to ford or run up and down creek beds is also an important handling and mobility feature.

Some manufacturers attend and sponsor off-road rallies. This gives them instant access to real world off-roading situations. They also get feedback from customers and gain information that can be used in setting up off-road durability testing.

A good test program can eliminate field problems. Weaknesses and shortcomings of a design can be found before going into production. This allows improvements and corrections to be made before offering the systems to the consumer.

SUMMARY AND CONCLUSIONS

The history of four-wheel-drive and all-wheel-drive vehicles will no doubt continue long into the future. However we will see changes in the vehicles, the drive systems and the way they are manufactured and used.

Commercial and military vehicles will probably remain unchanged in regard to their practical approach to four-wheel-drive. We can expect to see manually shifted part-time rugged systems in this category of vehicles for a long time.

The recent trend of people using a sport utility vehicle or pickup truck as a major part of their family transportation has certainly increased awareness of four-wheel-drive and all-wheel-drive. As some of these people return to or add a passenger car to their family fleet, they will want the added utility of all-wheel-drive.

In a world where we are very familiar with 2-wheel-drive vehicles, one would never propose a device to bypass the axle differential and drive only one wheel. Not even to save gasoline. Why complicate the system with disconnects, shift levers, push buttons, added cost and complexity when the potential gains are so small?

Yet we go on building 4-wheel-drive vehicles using the same convoluted logic! Most part-time 4-wheel-drive vehicles are shifted into the four-wheel-drive mode only a few times in their entire useful life! We build sophisticated systems to drive all the wheels, then add even more complexity *and cost,* to render them useless. Furthermore, the owner never gets to experience the pure joy of operating a full-time four-wheel-drive vehicle. She is deprived of the improved safety, handling, stability, cornering, mobility, and general feel that can be provided.

Full-time four-wheel-drive has been lumped into the category of just another feature, like a trim package or fancy wheel covers. *It is far more important than that!* The entire character of the vehicle is transformed. As the buying public becomes more aware of this difference, they will demand the extra performance and safety offered by a well engineered full-time four-wheel-drive system.

WHERE DO WE GO FROM HERE? — The ideal all-wheel-drive system is one in which torque to each wheel can be varied as required by performance objectives. Two ways of achieving this goal are:

- A continuously variable mechanical drive to each wheel.
- A hydraulic drive with pressure and flow proportioning to each wheel.

The challenge for the future all-wheel-drive designer is to create a system that achieves the torque proportioning objective, is simple, is economical to produce, and requires a minimum of maintenance.

ACKNOWLEDGMENTS

Over the years a great many people have shared their insight and knowledge of how mechanisms work and how to explain this vision to others. To all of you, I am grateful and hope that some of your insight will be transferred to still more people through this document.

In fear of an omission not everyone involved in this specific effort was listed here. Some of those who made a special contribution were as follows:

- Bernice I. Dick who helped with the research, organized and wrote much of the history, and typed, edited, and arranged the entire manuscript.
- Joseph A. Dick for being my technical advisor and contributor of ways to present information in a direct and understandable language.
- Ralph Baxter for carrying much of the department load at times when the writing consumed much of my time.
- Steve Goddard who helped with the proof reading and suggested some important inclusions.
- Dale Klein and Randy Sommer for contributing many of the figures and illustrations.
- Margie Foote for her efforts in proofreading and in helping to meet some serious deadlines.
- Rich Newer, of Chrysler Corporation, a contributor of technical information and illustrations.
- Jeff H. Skorupski and the Buckendale committee for their patience, confidence, and support of my efforts.
- Patricia Couhig for her help in logistics, detail, and especially in communications with the committee.
- John Hickey for his support of the total all-wheel-drive program.
- Dana Corporation and the Spicer Axle Division staff for their support of the effort.
- FWD Corporation for contributing the cover photograph and some of the technical material.
- Hagie Manufacturing Company for technical information on a unique four-wheel-drive system.
- Spicer Universal Joint Division for technical material.
- Eaton Corporation for technical material.

Also thanks to L. Ray Buckendale and the SAE for establishing the lecture series
and thanks to Dana Corporation, Eaton Corporation, and Rockwell International Corporation
for providing financial support for the lectures.

BIBLIOGRAPHY

Baxter, Ralph W. Jr. "A Four Wheel Drive Transfer Case with Internal VR Motors for Shifting and Clutch Actuation", SAE Technical Paper Series 933053, 1993.

Cameron, Dugald, et al. "1991 Model Year Chrysler Mini–Van All Wheel Drive Vehicle", SAE Technical Paper Series 901762, 1990.

Church, Austin H. *Guillet's Kinematics of Machines*. 5th ed. New York: John Wiley & Sons, Inc., 1956.

Clymer, Floyd. *Those Wonderful Old Automobiles*. New York: Bonanza Books, 1953.

CONVEL Constant Velocity Joints Brochure, CONVEL Inc., Summerville, South Carolina: January, 1994.

Crow, James T. & Spencer Murray. *Off-Roader's Handbook*. Los Angeles, California: HPBooks, 1986.

Deere & Company. *John Deere Tractors 1918-1987*. St Joseph, Michigan: American Society of Agricultural Engineers, 1987.

Faith, Nicholas. *Classic Trucks*. Osceola, Wisconsin: Motorbooks International Publishers & Wholesalers, 1995.

Goddard, Steve, "Vehicle Drawbar Characteristics of Electric and Viscous Clutch All Wheel Drive Systems", SAE Technical Paper Series 933052, 1993.

Gunnell, John G., ed. *American Work Trucks: A Pictorial History of Commercial Trucks 1900–1994*. Iola, Wisconsin: Krause Publications, 1993.

Gunnell, John G. *The Complete Four-Wheel-Drive Manual*. Blue Ridge, Pennsylvania: Tab Books, Inc., 1981.

Gunnell, John G. *100 Years of American Cars*. Iola, Wisconsin: Krause Publications, 1993.

Hermanns, Martin J. "Front Drive Systems for Four–Wheel Drive Light Trucks", SAE Publication SP–437, the 25th L. Ray Buckendale Lecture, 1979.

Karolevitz, Robert F. *This Was Trucking*. New York: Bonanza Books, 1966.

Mazziotti, Philip J. "Dynamic Characteristics of Truck Driveline Systems", SAE Publication SP–262, the 11th L. Ray Buckendale Lecture, 1964.

Montville, John B. *Mack*. Tucson, Arizona: AZTEX Corporation, 1979.

Nichols/Portland Brochure, W. H. Nichols Company, Nichols/Portland Division, Portland, Maine: undated.

Richardson, John and Dick, Wes. "Electronic Transfer Case for 1990 Aerostar Electronic Four Wheel Drive", SAE Technical Paper Series 892538, 1989.

Robson, Graham. *The Land Rover, Workhorse of the World*. North Pomfret, Vermont: David & Charles, Inc., 1977.

Spicer Axles Specifications, Dana Corporation, Spicer Axle Division, Fort Wayne, Indiana: October, 1987.

Swinford, Norm. *Allis–Chalmers Farm Equipment (1914-1985)*. St Joseph, Michigan: American Society of Agricultural Engineers, 1994.

Universal Joint Layout and Selection Data Book, Rockwell–Standard Corporation: 1964.

Vanderveen, Bart H., ed.. *American Trucks of the Early Thirties;* Oslager Auto Library, Great Britain: BAS Printers, 1974.

The World of Automobiles, An Illustrated Encyclopedia of the Motor Car Vol. 1-7. New York: Columbia, 1974.

APPENDIX – 1

CALCULATING PLANETARY OR EPICYCLIC GEAR REDUCTION RATIOS

INTRODUCTION

Planetary reduction gear sets are commonly used in two speed transfer cases. The following method allows for simple deduction of ratios instead of memorizing or referring to a multiplicity of complicated formulas. It is useful in understanding torque splits in planetary differentials.

The following is from a textbook studied by this author during his college days. In all the years since, I have never found a better system for calculating epicyclic gear ratios.

Excerpt From *Guillet's Kinematics of Machines*, 5th ed., pp. 220–227:

9.8 Second Method A general method of calculating the velocity ratio of epicyclic trains that may be used when no member is held fixed will now be described. It consists of three steps:

1. The whole train is locked so that there can be no relative motion of the parts, and it is then rotated one revolution clockwise. As a result each member of the train will rotate one revolution clockwise. This may be clarified by slowly rotating this book one revolution and noting that every member in a picture of a train makes one revolution about its own axis.

2. The epicyclic train is now converted into an ordinary train by locking the arm on which the gears are mounted and at the same time releasing the fixed gear. The gear formerly fixed is now rotated one turn counterclockwise, and the number of turns made by all the members recorded on a table.

3. The net result of the previous operations may be found by adding algebraically the number of revolutions made by each member of the train. The resultant movement of the "fixed" gear is always zero; hence the angular displacements of the other gears are the same as though the train had remained an epicyclic. From these angular displacements, the velocity ratio of the train can be calculated.

Example 1 Applying the above method to the epicyclic train of Fig. 9•8, we have: (1) the train is locked and the whole mechanism is rotated one revolution clockwise; (2) arm 4 is locked and gear 1 rotated one turn in a counterclockwise or negative direction, and the number of turns of the other gears are noted on the table; (3) the resultant motion of the members concerned is found by adding algebraically the values set up in the table for steps *1* and *2*. This is shown in table 9•1.

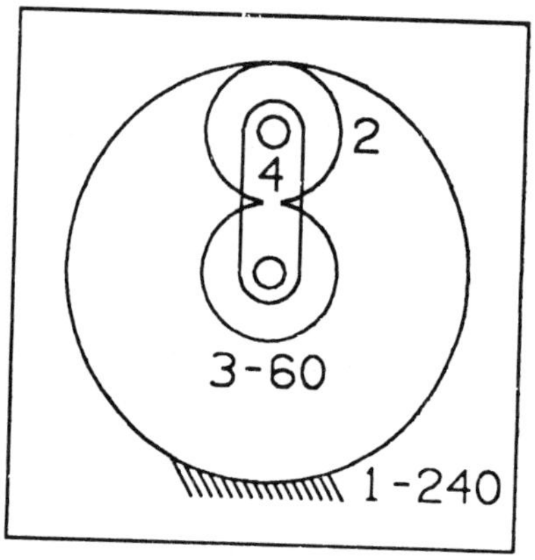

Fig. 9•8

Table 9•1
Turns

Step	*1*	*2*	*3*	*4*
1	+1	+1	+ 1	+1
2	– 1	...	+ 4	0
3	0	...	+5	+1

Thus while the arm 4 makes +1 turn, gear 3 makes +5 turns, and the ratio ω_{31}/ω_{41} equals +5.

Example 2 In Fig. 9•9 is illustrated a compound epicyclic train in which the arm 6 carrying gears 3 and 4 is neither the driver nor the driven link. Driver 2 engages gear 3, which in turn meshes with the stationary annular gear 1. Gears 3 and 4 are keyed to a shaft supported by arm 6, which is free to turn on shaft *A*. Gear 4 engages annular gear 5, which is keyed to the driven shaft *B*.

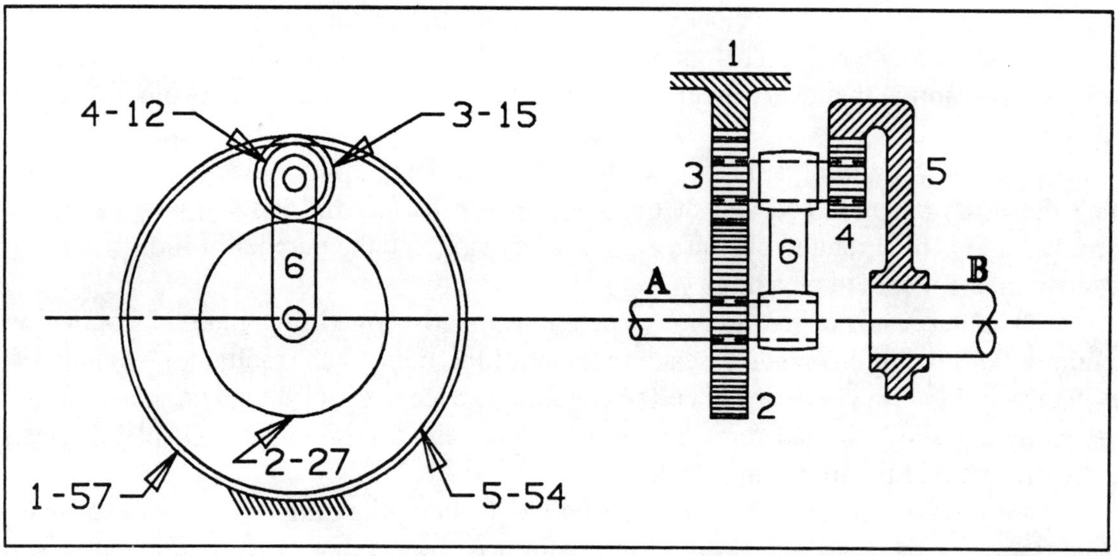

Fig. 9•9

The first step in finding the train ratio is to lock the train and rotate the mechanism one turn in a positive (clockwise) direction. The second step is to fix the arm **6**, rotate the fixed gear **1** one revolution in a negative (counterclockwise) direction, and tabulate the turns of the other gears. The third step is to add algebraically the motions obtained in the steps already taken to find the resultant motion. This is done in Table 9•2.

Table 9•2

Turns

Step	*1*	*2*	*3*	*4*	*5*	*6*
1	+1	+1	+1	+1	+1	+1
2	−1	$+\frac{57}{27}$	$-\frac{57}{15}$	$-\frac{57}{15}$	$-\frac{57}{15} \cdot \frac{12}{54}$	0
3	0	$+\frac{28}{9}$	...	...	$+\frac{7}{45}$	+1

Thus, while gear 2 makes +28/9 turns, gear 5 makes +7/45 turns. The speed ratio $\omega_{51}/\omega_{21} = \dfrac{7/45}{28/9} = +1/20$. Hence, while shaft *A* makes 20 turns, shaft *B* makes 1 turn in the same direction.

Example 3 Figure 9•10 shows a compound reverted epicyclic train arranged to give a large speed reduction from the driving shaft *A* to the driven shaft *C*. Shaft *A* drives an arm 5, which supports shaft *B* to which are keyed gear wheels 2 and 3. Wheel 1 is fixed. Rotation of *A* therefore causes 2 to roll on 1; 3 meanwhile driving 4, which is keyed to the driven shaft *C*. Suppose that the numbers of teeth on the gears are: 1—60 teeth, 2—61 teeth, 3—60 teeth, 4—61 teeth. shaft *A* turns 100 rpm clockwise. Find the speed and direction of the rotation of shaft *C*.

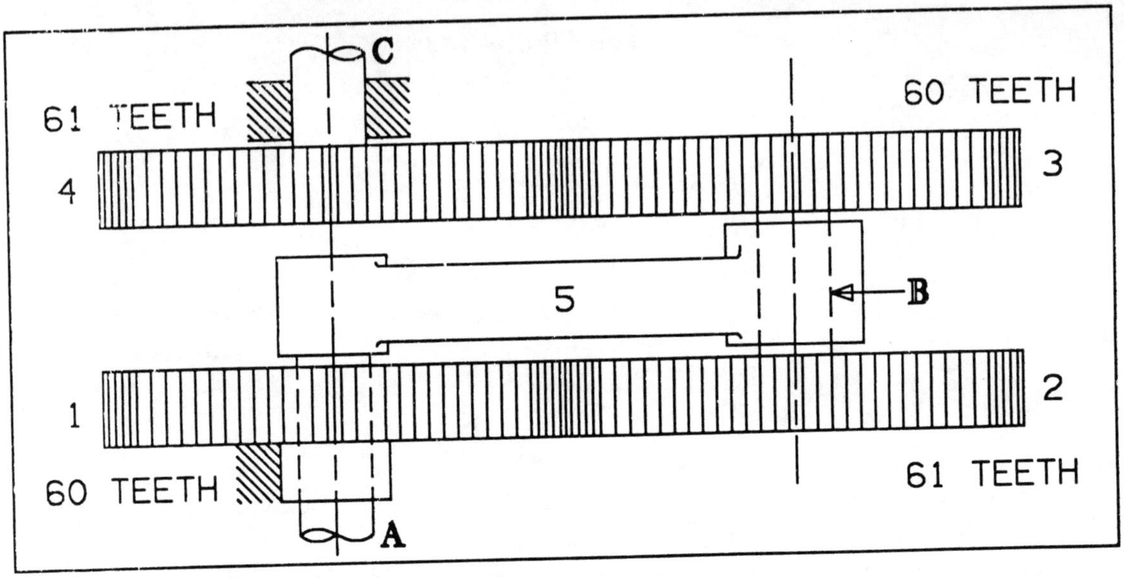

Fig. 9•10

Following the usual procedure, we first rotate the entire train one turn clockwise (+1). Then we hold the arm 5 fixed, rotate the fixed gear 1 one turn in the opposite direction, and find the number of turns made by the other gears, as shown in step *2* of the table. Finally the resultant motion is obtained in step *3* by adding steps *1* and *2* algebraically, as shown in Table 9•3. From this it may be seen that, if shaft *A* makes 100 rpm clockwise, *C* makes 100(121/3721) = 3.25 rpm in the same direction.

This case illustrates the method by which a large speed reduction can be obtained by means of an epicyclic train, using gears which are all about the same size. The train may thus be made in a compact form.

Table 9•3

Turns

Step	*1*	*2*	*3*	*4*	*5*
1	+1	+1	+1	+1	+1
2	−1	$+\frac{60}{61}$	$+\frac{60}{61}$	$-\frac{60}{61}\bullet\frac{60}{61}$	0
3	0	…	…	$+\frac{121}{3721}$	+1

It will be instructive to investigate this example further by interchanging the two pairs of gears. Then the numbers of teeth on the gears are: 1—61 teeth, 2—60 teeth, 3—61 teeth, 4—60 teeth. Shaft *A* again turns 100 rpm clockwise, and it is desired to find the speed and direction of rotation of shaft *C*.

The procedure is exactly the same as before, and is summarized in Table 9•4.

Table 9•4

Turns

Step	*1*	*2*	*3*	*4*	*5*
1	+1	+1	+1	+1	+1
2	− 1	$+\frac{61}{60}$	$+\frac{61}{60}$	$-\frac{61}{60}\bullet\frac{61}{60}$	0
3	0	…	…	$-\frac{121}{3600}$	+1

From this it may be seen that, if shaft *A* rotates 100 rpm clockwise, shaft *C* rotates 100 (121/3600) = 3.36 rpm in the opposite direction.

Hence, merely interchanging the two sets of gears is sufficient to cause the direction of rotation of the driven shaft to be reversed, the size of the train and the length of the arm 5 being unaltered.

9.9 Epicyclic trains with no gear fixed. Occasionally an epicyclic gear train may be used in which no gear is held fixed. The most common example of this is an automobile differential when the car is turning a corner, which will be considered later.

The procedure in problems of this type is similar to the second method outlined in Art. 9•8, except that in the first step, instead of the locked train being rotated one revolution, it is rotated as many revolutions as the arm makes per unit of time; and in the second step, when the arm is locked, the gear whose absolute speed is known must be rotated in the direction a number of times which, when added to the turns of the whole train in the first step will give the correct algebraic total. It is perhaps best illustrated by an example.

Example 1 In Fig. 9•11, shaft *A*, rotating 100 rpm clockwise, drives an epicyclic gear train by the meshing of gear 2 with 3, and 4 with 5. It should be realized that gears 2 and 4 on shaft *A* do not form part of the epicyclic train but merely drive it. The numbers of teeth on each gear are shown on the figure. It is desired to know the direction of rotation and speed of shaft *B*.

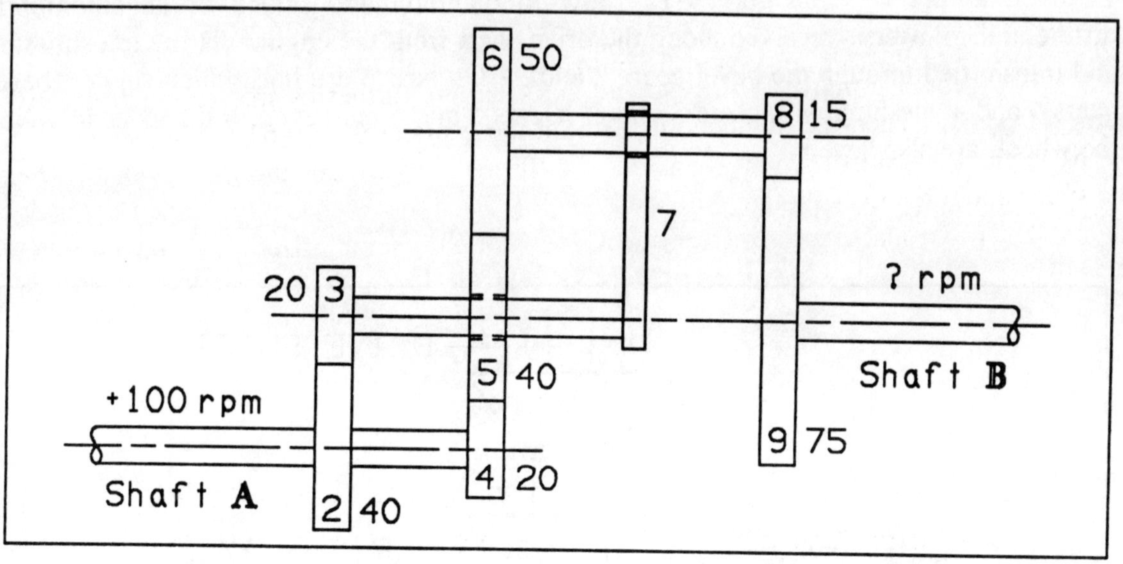

Fig. 9•11

The speed of gear 3, or the arm 7 which is keyed to 3, is $100 \times \dfrac{40}{20} = -200$ rpm.

The speed of gear 5 is $100 \times \dfrac{20}{40} = -50$. Table 9•5 may now be set up listing the gears in the epicyclic train.

Table 9•5

Turns

Step	*3 or 7*	*5*	*6*	*8*	*9*
1	– 200	– 200	– 200	– 200	– 200
2	0	+ 150	$-150 \cdot \frac{40}{50}$	$-150 \cdot \frac{40}{50}$	$+150 \cdot \frac{40}{50} \cdot \frac{15}{75}$
3	– 200	– 50	– 320	– 320	– 176

For step *1* the entire train is rotated 200 revolutions counterclockwise, which is the speed and direction of rotation of the arm 7. For step *2* the arm 7 is held fixed, and gear 5 revolved a number of turns in the proper direction so that its resultant motion will be that actually occurring, or –50 rpm. The required value is evidently +150 rpm. With the arm 7 fixed and gear 5 rotated +150 revolutions, the revolutions of the other gears may be found and listed. In step *3* the resultant motion is found by algebraic addition, and it is found that gear 9, or shaft *B*, rotates –176 rpm, that is, counterclockwise.

Example 2 Figure 9•12 shows a diagrammatic sketch of an automobile differential. Power is received along the drive shaft from the engine via the transmission and transmitted through the bevel gears 2 and 3. On gear 3 are lugs which support bevel gears 4 and 4' meshing with gears 5 and 6. Gears 5 and 6 are keyed to the axles to which the wheels are also keyed.

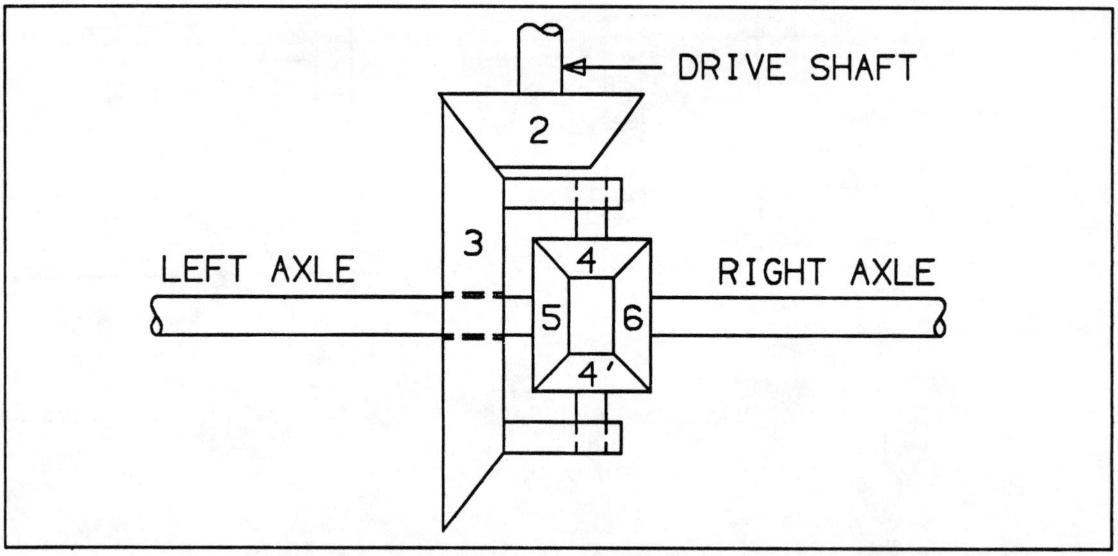

Fig. 9•12

When the car is traveling along a straight road, the differential turns "end over end"; that is, there is no relative motion between gears 4 or 4' and gears 5 and 6. However, when the auto turns a corner, one wheel and axle must travel slower and the other speed up to prevent slippage of the tire on the road. This causes relative motion of the gears in the epicyclic train of which 3 is the arm, and the other gears are 4 or 4', 5 and 6. Assume that on a straight road each axle turns 100 rpm, but that on a curve the speed of the right wheel drops to 50 rpm. What will be the speed of the left axle, assuming that the drive-shaft speed is unchanged?

Following the method outlined above (see Table 9•6), it is known that the speed of the gear or arm 3 is +100 rpm. Hence, in step *1* we rotate the train +100 revolutions. In step *2* the arm, or gear 3, is held fixed and gear 6 revolved a number of turns in the proper direction so that is resultant motion will be that actually occurring, or +50 rpm. This value is −50 turns. With the arm 3 fixed and gear 6 rotated −50 turns, it is apparent that gear 5 will rotate +50 turns. This is true since gears 5 and 6 are the same size, and gears 4 and 4' merely act as idlers. When the values are entered in the table, the resultant motion is found in step 3. It may be observed that the speed of one wheel will increase the same amount that the other decreases. If the rear end is jacked up and one wheel held fixed, the other rotates twice as fast. This situation occurs frequently in the winter when one wheel of an auto is resting on ice having a low coefficient of friction and the other is on a dry road having a high coefficient of friction. In such cases the car must be pushed away from this condition before it can be driven.

TABLE 9•6

Turns

Step	*3*	*5*	*6*
1	+100	+100	+100
2	0	+50	− 50
3	+100	+150	+50

APPENDIX – 2

SAE recommended practice for communicating and depicting all-wheel-drive and four-wheel-drive systems is contained in specifications:

- J1952 All–Wheel–Drive Systems Classification
- J2059 All–Wheel–Drive Drivetrain Schematic Symbols Standards

These specifications are to be found in the SAE Handbook and were printed in the publication "Introduction to All–Wheel–Drive", SAE HS–2300, September, 1992.